AF845916

Nitroimidazoles

Chemistry, Pharmacology, and Clinical Application

NATO ADVANCED STUDY INSTITUTES SERIES

A series of edited volumes comprising multifaceted studies of contemporary scientific issues by some of the best scientific minds in the world, assembled in cooperation with NATO Scientific Affairs Division.

Series A: Life Sciences

Recent Volumes in this Series

Volume 37 – Active Defense Mechanisms in Plants
edited by R. K. S. Wood

Volume 38 – Cell Growth
edited by Claudio Nicolini

Volume 39 – Drug Development, Regulatory Assessment, and Postmarketing Surveillance
edited by William M. Wardell and Giampaolo Velo

Volume 40 – Chromosome Damage and Repair
edited by Erling Seeberg and Kjell Kleppe

Volume 41 – Protein Biosynthesis in Eukaryotes
edited by R. Pérez-Bercoff

Volume 42 – Nitroimidazoles: Chemistry, Pharmacology, and Clinical Application
edited by A. Breccia, B. Cavalleri, and G. E. Adams

Volume 43 – Advanced Topics on Radiosensitizers of Hypoxic Cells
edited by A. Breccia, C. Rimondi, and G. E. Adams

Volume 44 – Cell Regulation by Intracellular Signals
edited by Stéphane Swillens and Jacques E. Dumont

Volume 45 – Structural Molecular Biology: Methods and Applications
edited by David B. Davies, Wolfram Saenger, and Stephen S. Danyluk

Volume 46 – Post-Harvest Physiology and Crop Preservation
edited by Morris Lieberman

This series is published by an international board of publishers in conjunction with NATO Scientific Affairs Division

A Life Sciences B Physics	Plenum Publishing Corporation London and New York
C Mathematical and Physical Sciences	D. Reidel Publishing Company Dordrecht, The Netherlands and Hingham, Massachusetts, USA
D Behavioral and Social Sciences E Applied Sciences	Martinus Nijhoff Publishers The Hague, The Netherlands

Nitroimidazoles

Chemistry, Pharmacology, and Clinical Application

Edited by

A. Breccia
University of Bologna
Bologna, Italy

B. Cavalleri
Gruppo Lepetit SpA
Milan, Italy

G. E. Adams
Institute of Cancer Research
Sutton, England

PLENUM PRESS • NEW YORK AND LONDON
Published in cooperation with NATO Scientific Affairs Division

Library of Congress Cataloging in Publication Data

NATO Advanced Study Institute on Radiosensitizers of Hypoxic Cells (1980: Cesenatico, Italy)
Nitroimidazoles: chemistry, pharmacology, and clinical applications.

(NATO advanced study institutes series. Series A, Life sciences; v. 42)
"Proceedings of an international confeerence on nitroimidazoles: chemistry, pharmacology, and clinical application, held August 27–30, 1980, in Cesenatico, Italy. The conference was the first part of the NATO Advanced Study Institute on Radiosensitizers of Hypoxic Cells, held August 27–September 7, 1980, in Cesenatico, Italy"—T.p. verso.
"Published in cooperation with NATO Scientific Affairs Division."
Includes bibliographies and index.
1. Nitroimidazoles—Physiological effect—Congresses. 2. Nitroimidazoles—Testing—Congresses. 3. Cancer—Adjuvant treatment—Congresses. I. Breccia. A. II. Cavalleri, B. III. Adams, Gerald E. IV. North Atlantic Treaty Organization. Division of Scientific Affairs. V. Title. VI. Series. [DNLM: 1. Nitroimidazoles—Pharmacodynamics—Congresses. 2. Nitroimidazoles—Therapeutic use—Congresses. QU 65 N731 1980]
RM666.N727N37 1980 615.7 82-610
ISBN 978-1-4684-4153-6 ISBN 978-1-4684-4151-2 (eBook) AACR2
DOI 10.1007/978-1-4684-4151-2

Proceedings of an international conference on Nitroimidazoles: Chemistry, Pharmacology, and Clinical Application, held August 27–30, 1980, in Cesenatico, Italy. The conference was the first part of the NATO Advanced Study Institute on Radiosensitizers of Hypoxic Cells, held August 27–September 7, 1980, in Cesenatico, Italy.

© 1982 Plenum Press, New York
Softcover reprint of the hardcover 1st edition 1982
A Division of Plenum Publishing Corporation
233 Spring Street, New York, N.Y. 10013

All rights reserved

No part of this book may be reproduced, stored in a retrieval system, or transmitted, in any form or by any means, electronic, mechanical, photocopying, microfilming, recording, or otherwise, without written permission from the Publisher

FOREWORD

In July 1979, the Faculty of Pharmacy of the University of Bologna received a proposal from Professor Breccia to hold an International Conference on "Nitroimidazoles: Chemistry, Pharmacology and Clinical Application".

Because of the great interest in these drugs in various fields, I was very pleased to accept the proposal and to give to the Conference, the sponsorship of the University of Bologna. It was an added pleasure to accept the chairmanship of the meeting, together with Professors Sensi and Adams.

At the same time, the Minister of Education approved the proposal also from the Faculty of Pharmacy, to offer an advanced Course on the topic: "Radiosensitizers of Hypoxic Cells". This Course was subsequently approved by a special Committee of the NATO Scientific Programme as an Advanced Study Institute. Since the subject matter of both the Conference and the Course are closely inter-related, it was logical to plan the Conference as part of the overall Course.

Bringing together specialists from quite different and apparently unrelated fields of research, but all with an interest in the study and application of one single group of drugs, is a novel idea with intriguing possibilities. The volumes of the proceedings include contributions from experts with such diverse interests as synthetic chemistry, mechanism of drug action, parasitology, anaerobic bacteria and protozoa, pharmacology and toxicology, radiation sensitizers for use in radiotherapy and the development of drugs for use in cancer chemotherapy generally.

I would like to congratulate the members of the Scientific Committees on the manner in which they have designed the scientific programme. I would also like to thank the Committees, the speakers, chairmen and coordinators for so successfully bringing together such a high standard of scientific and clinical experience from many different countries of the world.

I would like to say a few words about the region in which the meetings took place. Cesenatico is part of the Romagna which is one of the most famous of all the Italian Rivieri. It is, however, not just a holiday area but a scientific one also. As some of you may know, one of the most advanced University Centres of Marine Biology is located in this town. Conference initiatives such as this one are fairly frequent, for example a Meeting on Polyammines in Cancer Research will take place soon in nearby Rimini.

I am sure that the interest in the topics in these proceedings will not be limited only to the information about research already carried out but will be the basis for a better development of new drugs in cancer therapy.

Professor Carlo Rizzoli
Rector
University of Bologna
Italy

PREFACE

The Advanced Study Institute sponsored by NATO, on Radiosensitizers of Hypoxic Cells, was divided into two sections. The first section of the Institute consisted of an International Conference on Nitroimidazoles which brought together a broad range of interests in the chemistry and pharmacology of these compounds and in their application in various fields in chemical medicine.

The second section was concerned with a series of papers on the general subject of Radiosensitizers of Hypoxic Cells. It included lectures and workshops on the chemical, biological and clinical developments in this field. For convenience, and also because of the large number of papers presented, the total proceedings of the Institute are published in two volumes.

A. Breccia

B. Cavelleri

G.E. Adams

ACKNOWLEDGEMENTS

The Organizers gratefully acknowledge the following for their generous financial support, including fellowships to some participants:

A.S.I. - N.A.T.O.

Azienda di Soggiorno e Turismo di Cesenatico

Banca del Monte di Bologna

Camera di Commercio di Forli

Istituto De Angeli

Lega Nazionale lotta contro i Tumori sez. di Bologna

Ministero della Pubblica Istruzione

Ospedale "M. Malpighi", Bologna

Provincia di Forli

Regione Emilia-Romagna

Società Carlo Erba Farmitalia

Società Cesenate Corse al Trotto

Società Elettronica, Roma

Società Farmigea

Società Hoffmann-La Roche

CONTENTS

THE PHARMACOLOGY AND TOXICOLOGY OF 5-NITROIMIDAZOLES

G.L. Biagi

Istituto di Farmacologia
Università di Bologna
Italy

INTRODUCTION

The nitroimidazoles belong to the group of nitroheterocyclic compounds which play an important role as drugs in chemotherapy. The first heterocyclic compound to be used in human medicine was nitrofurazone, a nitrofuran derivative employed during the war for its antibacterial properties for the topical treatment of burns and wounds (1). The discovery of the antibacterial properties of the nitrofurans stimulated interest in nitroheterocyclic compounds as chemotherapeutic agents. As a consequence, several series of nitroheterocyclic compounds were studied. In particular, nitroimidazoles gained importance because of their antiprotozoal activity. The purpose of this paper is to provide a basic introduction to the pharmacology and toxicology of nitroimidazoles as an aid in the study of more detailed aspects which will be covered in the other papers in this Conference on nitroimidazoles.

Finally, when dealing with the structure-activity relationships of 5-nitroimidazoles some data describing our own work will be presented.

PHARMACOLOGICAL EFFECTS AND THERAPEUTIC USES

Antitrichomonal activity (2)

The discovery of the antiparasitic activity of azomycin (2-nitroimidazole) has resulted in many compounds possessing interesting biological activity. Workers at Rhône-Poulenc discovered metronidazole which is now the drug of choice for the treatment of human trichomoniasis. A series of investigations at

Merck resulted in the discovery of very active 2-substituted 1-methyl-5-nitroimidazoles such as flunidazole, "MF" nitromidazole, ronidazole.

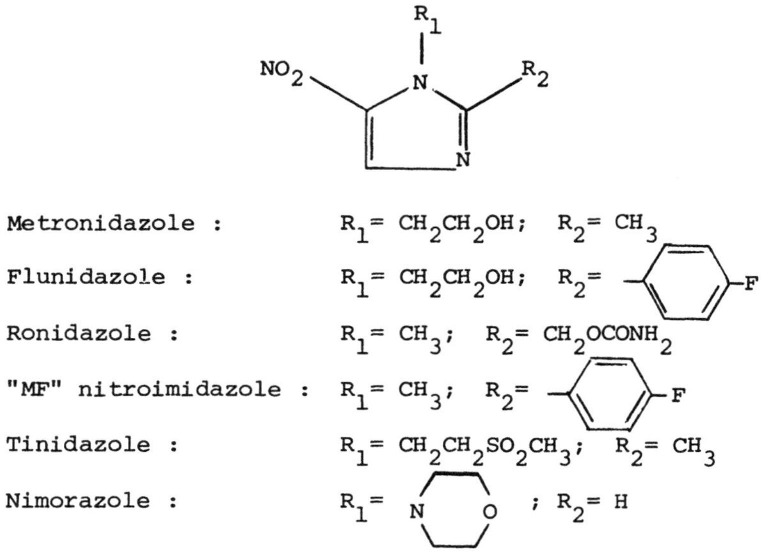

The simplest structural requirement for biological activity is the 1-alkyl-5-nitroimidazole unit. Moreover the nitro group should not be sterically crowded. Development of this theme resulted in the discovery of tinidazole and nimorazole as effective drugs for human trichomoniasis.

Antiamoebic activity (2)

Laboratory tests showed that metronidazole was amoebicidal *in vitro* and effective against intestinal amoebiasis in rats and hepatic amoebiasis in hamsters. The initial clinical trials indicated that it was effective in invasive amoebic dysentery and amoebic liver abscess. Subsequently metronidazole has been established as the drug of choice in the treatment of all forms of amoebiasis.

Among the metronidazole derivatives, panidazole, tinidazole, nimorazole, flunidazole, ronidazole and "MCA" nitroimidazole were found to be effective as antiamoebic drugs. In particular, panidazole has an activity comparable to that of metronidazole against hepatic amoebiasis in hamsters. Tinidazole is active in amoebiasis in rats, hamsters, and humans. Nimorazole is active against intestinal and hepatic amoebiasis in humans.

However none of these compounds was shown to be superior to metronidazole.

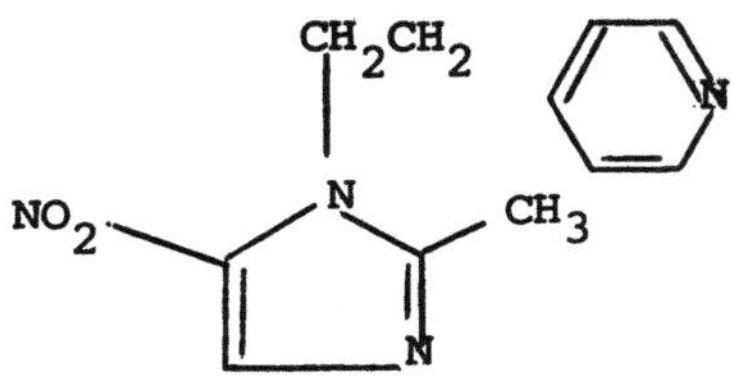

Panidazole

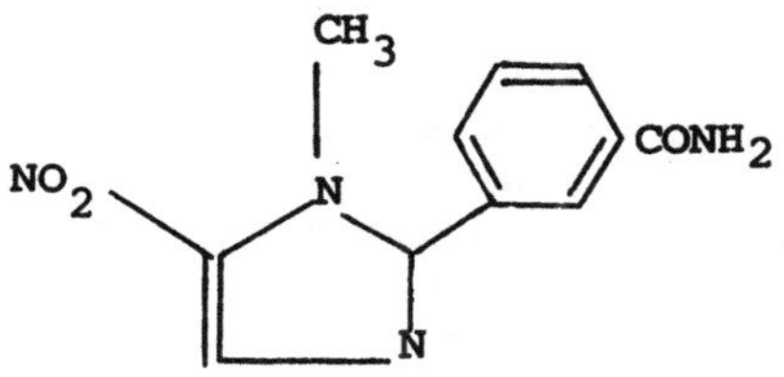

"MCA" nitroimidazole

Antigiardial activity (2)

Metronidazole has been shown to be effective in treating giardiasis, which is an infection of the intestinal tract of man caused by a protozoan flagellate, Giardia lamblia. Nimorazole is equally effective. The clinical effectiveness and the lower toxicity of metronidazole and nimorazole in giardiasis therapy presages their replacement of quinacrine.

Antihistomonal activity (2)

Histomoniasis is a disease of poultry, particularly of turkeys, caused by Histomonas meleagridis. The first really effective agent used for the control of Histomonas was 2-amino-nitrothiazole. The discovery of the prophylactic activity of dimetridazole led to the study of a new series of 5-nitroimidazole derivatives among which ipronidazole and ronidazole proved to be most effective.

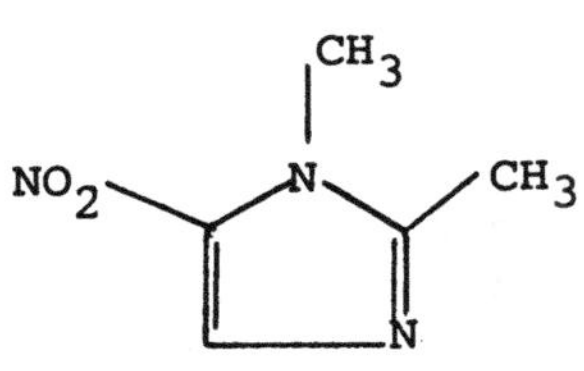

Dimetridazole

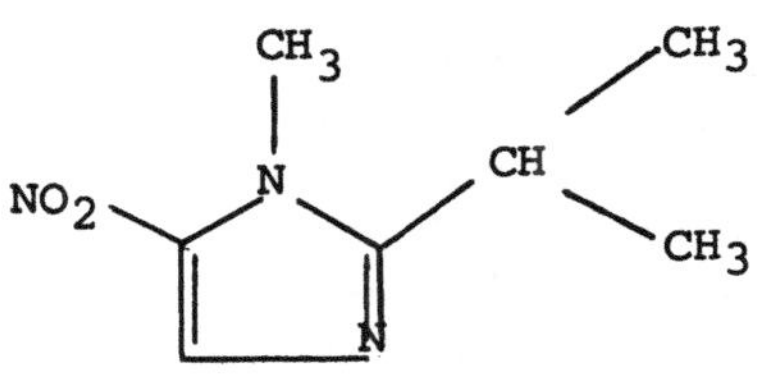

Ipronidazole

Antitrypanosomiasis and antidracontiasis activity (3,4)

Although the 5-nitroimidazoles do not play any important role in the treatment of human trypanosomiasis, compounds such as flunidazole, "MCA" nitroimidazole and ronidazole show also some antitripanosomial activity which seems to be due to inactivation of glycolytic enzymes. Finally metronidazole has been used with varying degrees of success in the treatment of dracontiasis. This is a nematode disease caused by Dracunculus medinensis. The mode of action of the 5-nitroimidazoles as antihelmintic agents is not clear. It might be due to an anti-inflammatory effect rather than a direct antihelmintic effect.

Antibacterial activity (1)

Investigations showed that nitroimidazoles also are active as antibacterial agents. In particular metronidazole and ronidazole were shown to be active against Gram-negative organisms. However toxicological studies revealed mutagenic or carcinogenic activity for many of these compounds. Therefore, since the mutagenic and antibacterial activity are intimately related at the molecular level, it does not seem likely that important non-mutagenic antibacterial nitroimidazoles will be found. As a consequence at the present moment the clinical status of metronidazole and ronidazole as antibacterial agents is not yet clear.

ABSORPTION, FATE AND EXCRETION

Among the 5-nitroimidazoles, metronidazole is the compound most studied from a pharmacokinetic point of view (4). The absorption after oral administration is usually very good and low concentrations of metronidazole appear in the saliva and in the breast milk during treatment. The metabolites result from oxidation of side chains and glucuronide formation. Both metronidazole and metabolites are excreted in the urine. The presence of pigments derived from the drug, makes the urine of some patients reddish-brown in colour.

TOXIC EFFECTS

The toxic effects induced by metronidazole can be taken as a guide to the study of the toxicology of other 5-nitroimidazoles(4). The most common side effects involve the gastrointestinal tract: nausea, anorexia, diarrhea, epigastric distress, abdominal cramping, metallic taste, dryness of the mouth. Vomiting is rare. Furry tongue, glossitis and stomatitis may be due to a sudden intensification of moniliasis.

As regards the CNS, dizziness, vertigo, parasthesia of an extremity and, very rarely, ataxia have been reported. Treatment should be discontinued promptly if ataxia or any other symptoms of CNS involvement occurs. Moreover metronidazole is contra-indicated in patients with active disease of the CNS. In the genitourinary tract, dysuria, cystitis, a sense of pelvic pressure, dryness of the vagina or vulva have been observed. Orticaria, flushing and pruritus are not rare.

While related chemicals have caused blood dyscrasia, serious difficulties have not been reported with metronidazole. In all cases, the white-cell count returned to normal after the course of medication was completed. However metronidazole is contra-indicated in patients with evidence or a history of blood dyscrasia. In some individuals the consumption of alcoholic beverages with metronidazole may produce a disulfiram-like effect. Metronidazole does not affect either the cardiovascular system or the respiratory apparatus. Metronidazole passes rapidly into the foetal circulation. However there is no evidence that its administration during pregnancy is responsible for foetal abnormality, prematurity or post-natal incident.

Carcinogenic effects due to the Medical use of metronidazole have not yet been reported (5). However the question may be raised whether the nitroimidazoles are indeed so harmless as generally supposed. In fact daily quantities of 2000 mg sometimes used in the treatment of certain forms of amoebiasis may cause serum levels as high as 30-45 μg/ml corresponding to a potential risk of increase of the mutation rate by a factor of 5 to 10.

It is also noteworthy that metronidazole increases the incidence of lung cancer in mice and that nothing is known about the potential mutagenic effect of the metabolites of metronidazole that appear in human urine.

STRUCTURE-ACTIVITY RELATIONSHIPS

Quantitative structure-activity relationship studies (QSAR) pointed out the importance of physico-chemical parameters in determining the biological activity of nitroimidazoles. In particular the importance of the lipophilic character and electro-negative groups in the R_1 position for the antischistosomal and antiprotozoal activity have been discussed (6,7). A relationship between antibacterial activity and electro-reduction has been pointed out by Chien et al (8). The frontier electron-density was shown to play a role in protein binding of metronidazole derivatives (9).

At the present time we are investigating the antibacterial and mutagenic activity of a series of 5-nitroimidazoles. The purpose of our work is to correlate activity with physico-chemical parameters such as the partition coefficient, the determination of which is in progress in our laboratory. Therefore, only very preliminary data are presented in this paper on nitroimidazoles. In Table 1 data are reported describing the antibacterial and mutagenic activity of some of the compounds we are investigating.

TABLE I

Compound	Mutagenic activity $\log \frac{1}{Cm}$ a	Antibacterial activity $\log \frac{1}{Ca}$ b	log P
1) Azanidazole	7.387	4.439	0.85
2) Ronidazole	4.881	2.074	-0.38
3) 1-Methyl-2-formyl-5--nitroimidazole	4.441	1.519	0.83
4) Ipronidazole	4.358	1.699	1.06
5) Carnidazole	4.058	1.561	0.90
6) 1-Methyl-2-hidroxymethyl-5--nitroimidazole	4.146	1.294	-0.03
7) Ornidazole	4.022	1.247	0.60
8) Tinidazole	4.183	1.210	-0.36
9) Metronidazole	3.733	0.874	-0.10
10) Nimorazole	3.735	0.880	0.07
11) 5-Nitroimidazole	0.863		-0.16
12) 2-Methyl-5-nitroimidazole	1.104		0.49

a = Cm is the concentration (μM/l) increasing 5x spontaneous retromutations;

b = Ca is the concentration (μM/l) determining an inhibitory zone of 30 mm diameter on solid saline media.

The relationship between mutagenic and antibacterial activity is described by equation 1.

$$\log \frac{1}{Cm} = 2.772 + 1.025 \log \frac{1}{Ca} \qquad 1$$

$$(n= 10;\ r= 0.991)$$

where Cm and Ca are the molar concentrations producing the biological responses described in Table 1, n is the number of data points and r is the correlation coefficient.

The highly significant (F = 450.9) linear relationship between mutagenic and antibacterial activity as expressed by eqn.1, supports the finding that the mutagenic and antibacterial modes of action are intimately related at the molecular level (1). The log P values, i.e. the partition coefficients in an octanol/water system, reported in Table 1 do not appear to correlate with either type of activity.

Compounds 11 and 12, which are among the most hydrophilic ones, have a very low mutagenic activity when compared with compounds 2 and 8, both of which are hydrophilic but show a much higher mutagenic activity. Therefore other factors must be taken into consideration in order to explain the mutagenic and antibacterial activity of 5-nitroimidazoles. The presence of a SO_2 group and/or H bonding could play an important role.

REFERENCES

1. E.R. Bambury, in: "Burger's Medicinal Chemistry", M.E. Wolff, ed., Part II, John Wiley & Sons, New York (1979).
2. W.J. Ross, in: "Burger's Medicinal Chemistry", M.E. Wolff, ed., Part II, John Wiley & Sons, New York (1979).
3. P.J. Islip, in: "Burger's Medicinal Chemistry", M.E. Wolff, ed., Part II, John Wiley & Sons, New York (1979).
4. L.S. Goodman and A. Gilman, "The Pharmacological Basis of Therapeutics", MacMillan Publishing Co., New York (1975).
5. C.E. Woogd, J.J. Van Der Stel and J.J.J.A.A. Jacobs, Mutation Res. 26:483 (1974).
6. Y. Lin, P.B. Hulbert, E. Bueding and C.H. Robinson, J. Med. Chem. 17:835 (1974).
7. M.W. Miller, H.L. Howes, Jr., R.V. Kasubick and A.R. English, J. Med. Chem. 13:849 (1970).
8. Y.W. Chien and S.Ś. Mizuba, J.Med.Chem. 21:374 (1978).
9. D.R. Sanvordeker, Y.W. Chien, T.K. Lin and H.J. Lambert, J. Pharm. Sci. 11:1797 (1975).

NITROIMIDAZOLE CHEMISTRY I: SYNTHETIC METHODS

Bruno Cavalleri

Via Durando 38, 20158 Milano, Italy
Research Laboratories, Gruppo Lepetit S.p.A.

The first report (1) on the antitrichomonas activity of Azomycin*(I), obtained from fermentation of a strain of Nocardia mesenterica (2,3), appeared in 1956, and the first publications on synthetics (4) dealt with some transformations of natural 2-nitro-imidazole.

Although nitration of imidazoles had been reported quite a long time ago (5,6), a patent application for the first compounds synthesized with the aim of finding drugs with antitrichomonas activity was filed (7) in 1957 and the synthesis published in 1961 (7) and in 1966 (8). Metronidazole**(II) was among these compounds.

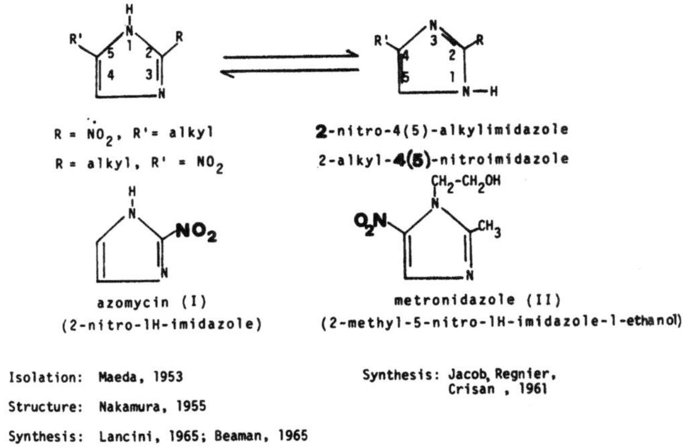

* 2-nitro-1H-imidazole.

** 2-methyl-5-nitro-1H-imidazole-1-ethanol.

Since then, the synthesis of 2-, 4- and 5-nitroimidazoles has undergone considerable development. Dinitro- (9,10) and halogeno-nitroimidazoles (11), which have been studied relatively little, have recently received new attention as potential radiosensitizers. Research efforts have produced several drugs that are in use today in the clinic or are being evaluated.

As the study of the relationships between the biological activity and the physico-chemical properties developed, some lines to pursue in the synthesis of new compounds became apparent. Some of the basic problems in the chemistry of the nitroimidazoles, especially those related to the existence of tautomeric forms (i.e. prototropic forms) of the compounds that do not have nitrogen substituents, were outlined by Hofmann (12).

Since 1963 many thousands of compounds have been synthesized, with 28 applications concerning 2-nitroimidazoles, mostly obtained by Hoffman-La Roche and Lepetit. For the 4- and 5-nitroimidazoles, the period of intensive development was between 1970 and 1975, and 314 patients had been filed, 65 of them by Merck.

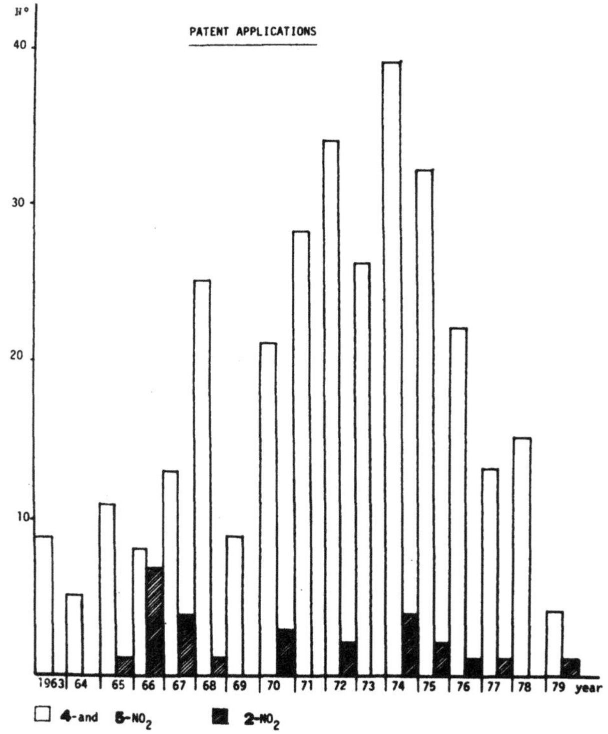

SCHEME I

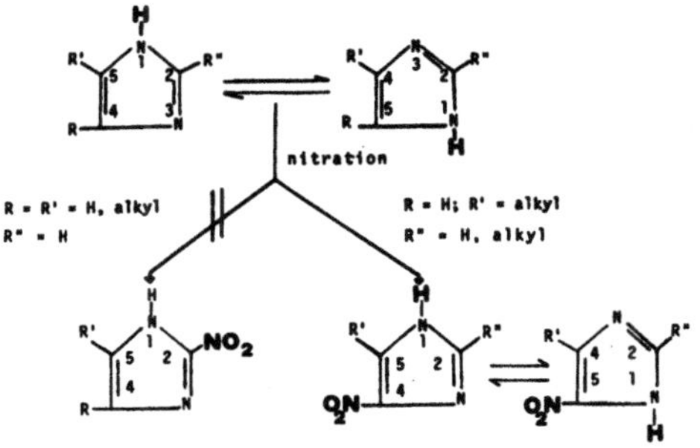

I. 2-NITROIMIDAZOLES

A. Introduction of the Nitro Group

It is well known that direct nitration of imidazole and of some of its derivatives leads to the introduction of the nitro group into either position 4 or 5, when these are free, and that there is no electrophilic substitution into position 2 even when 4 and 5 are occupied.

B. Biological Conversion of 2-aminoimidazoles

Lancini et al. (13) established that the addition of 2-aminoimidazole (III; R,R',R"=H) to cultures of an azomycin-producing *Streptomyces* strain resulted in increased yield of azomycin (I).

It was then demonstrated by Seki et al. (14) that 2-aminoimidazole (III) was accumulated, partly in the culture medium and partly in the mycelium, during fermentation of *Streptomyces eurocidicus*, and that this was converted to 2-nitroimidazole by washed cell suspensions. The microbial transformation of 2-aminoimidazoles was further investigated (15) (Table 1): Substrates carrying an alkyl group in the 4(5)-position were oxidized, the efficiency decreasing with the length of the chain (from methyl to *n*-butyl). 4,5-Dimethylimidazole gave lower yields, while no transformation was detected when there was a methyl group in position 1 or a phenyl in 4(5)-position. Substrate concentration also has a considerable effect.

TABLE I

BIOLOGICAL OXIDATION OF 2-AMINOIMIDAZOLES

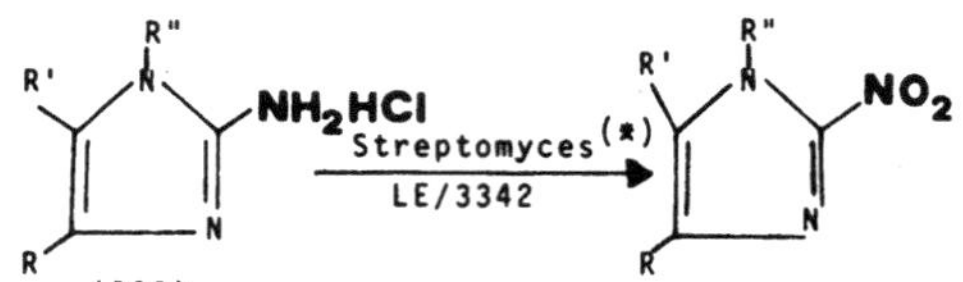

(III)

R	R'	R"	yield(%)
H	H	H	31.6 (azomycin, I)
CH_3	H	H	35.6
C_2H_5	H	H	38
$\underline{n}C_3H_7$	H	H	30
$\underline{i}C_3H_7$	H	H	5.8
$\underline{n}C_4H_9$	H	H	7
CH_3	CH_3	H	8.4
	H	H	no transformation
H	H	CH_3	no transformation

(*) growing cultures or washed mycelium suspension

Lancini et al. (1966,1968)

C. Chemical Conversion of 2-aminoimidazoles

The 2-aminoimidazoles, approaches to whose syntheses will be discussed later, are also the intermediates for chemical synthesis of 2-nitroimidazoles. There are only a few examples in the literature of the oxidation of an amino group in position 2 of the imidazole ring.

1-Benzyl-2-aminobenzimidazole (IV) can be transformed into the corresponding 2-nitrobenzimidazole (V) by treating it with liquid ammonia in the presence of metallic sodium (16), but this method does not apply to 2-aminoimidazoles (Scheme II; Eq. 1).

2-Aminoimidazoles cannot be oxidized by peroxytrifluoroacetic acid, which gives a good yield with some other heterocyclic rings, because imidazole has the amino group on a carbon atom adjacent to the two heteroatoms (17) (Eq. 2, 3).

SCHEME II

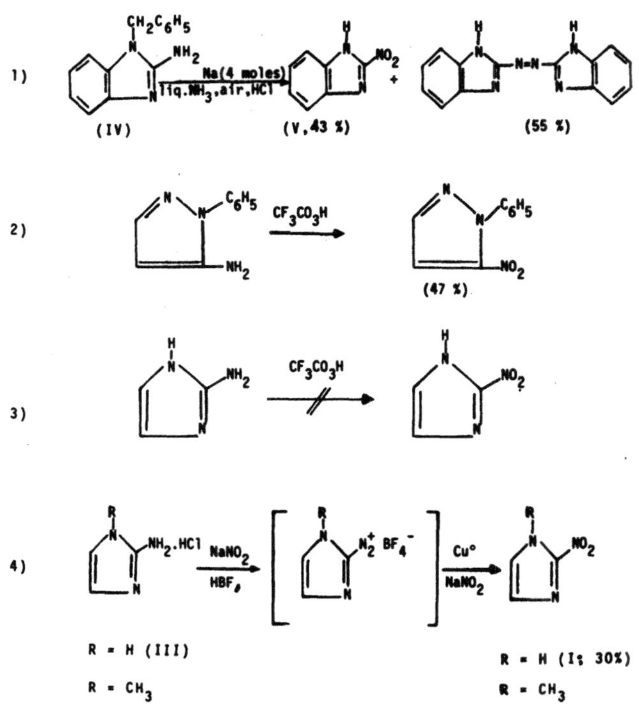

The transformation of the 2-amino group on the imidazole into a 2-nitro has been achieved through diazotization followed by the Gattermann reaction with sodium nitrite. Azomycin (I), and its N-methyl derivative, have been obtained by reacting the corresponding 2-aminoimidazoles at low temperature in fluoboric acid with sodium nitrite and treating the solution of the diazo intermediates with an excess of sodium nitrite in the presence of copper powder (18)(Eq. 4).

The same procedure has been applied to a series of 2-aminoimidazoles with different substituents, in sulfuric acid and in the presence of cupric sulfate (19). Both of these methods, with minor modifications, have been applied until recently to the synthesis of a large number of variously substituted 2-nitroimidazoles, with yields varying from 20 to 50% (Table 2).

The same reaction has also been applied to the preparation of compounds labelled with ^{14}C at the 2-C of imidazole ring (26).

When carbons 4 and 5 of 2-aminoimidazole carry two cyano groups, the corresponding 2-diazoimidazole (VII), although highly shock sensitive, was isolated (27). Treatment with sodium nitrite

TABLE II

CHEMICAL CONVERSION OF 2-AMINOIMIDAZOLES

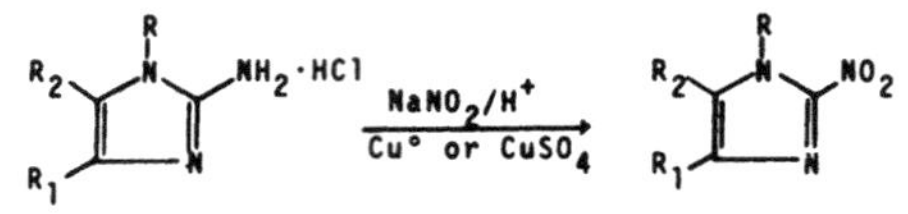

R	R_1	R_2	lit.
H	H	H	18,19
CH_3	H	H	18,19
H	$H(CH_3)$	$CH_3(H)$	20
H	CH_3	CH_3	20
CH_3 or C_2H_5	H	C_nH_{2n+1}	21
CH_3, C_2H_5	H	$CH_2CH_2X(X{=}OH,Cl)$	21,22
CH_3	H	$CH_2CHOH{-}CH_3$	21
$CH_2COOC_2H_5$	H	CH_3 or C_2H_5	21
CH_3	H	$COOR_3(R_3{=}CH_3,C_2H_5)$	23,22
CH_3	H	$(CH_3)_2C(OCH_3){-}$	24
CH_3	H	$CH_3{-}CH(CH_2OH){-}$	25

led to the 4,5-dicyano-2-nitro-imidazole (VIII), recovered as the tetraethylammonium salt (28) (Scheme III; Eq. 5).

A few examples have been given of the introduction of the nitro group by procedures different from those described above. The oxidation of the nitro group in position 2 by means of persulfuric acid has been reported (29) for 2-amino-4,5-dicarboxyimidazole (IX) (Eq. 6).

By treating 4(5)-methyl-5(4)-formylimidazole oxime (XI) with nitrogen tetroxide in acetonitrile the introduction of the nitro group in position 2 occurs (30), together with the transformation of the formyloximic group into a trinitromethyl (Eq. 7). Nitrogen tetroxide has also been applied (31) (Scheme IV; Eq. 8) to the 2-lithium derivative of 1-methyl-(or phenyl) imidazole (XIV).

1-Methyl-5-isopropyl-2- nitroimidazole (XIX) has been prepared (32) (Eq. 9) by a similar procedure, starting from the corresponding 2-bromoimidazole (XVIII). The method of synthesis of the latter intermediate from the dinitrile (XVI) is also interesting.

SCHEME III

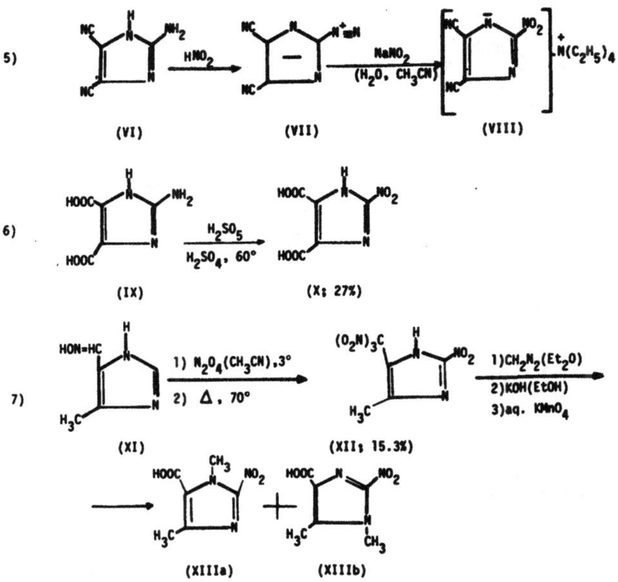

SCHEME IV

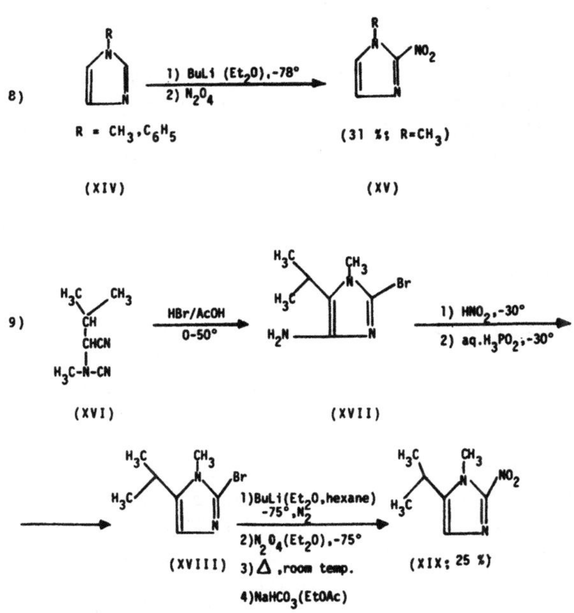

SCHEME V

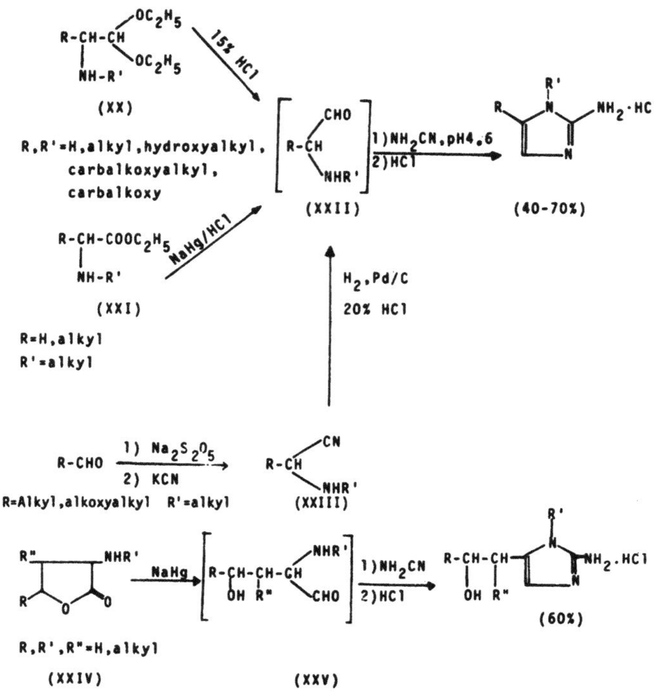

Recent publications show that the nitrogen tetroxide route can be widely applied, even with use of suitable protecting groups on the ring nitrogen (33). Recently (34), 2-nitroimidazoles have been obtained in 40÷70% yields by treating the lithium salts of N-trityl derivatives of 4-substituted (CH_3,CH_2OH) imidazoles with n-propyl nitrate.

D. Synthesis of 2-aminoimidazoles

Alkyl- and aryl-2-aminoimidazoles can be obtained by condensation of cyanamide with α-aminocarbonyl compounds (15,35,36). In detail (Scheme V), α-amino- and α-alkylaminoaldehydes (XXII), prepared by dilute acid hydrolysis of the corresponding acetals (XX) (21,22) or by Akabori reduction of suitable α-alkylaminocarboxylic esters (XXI)(21), are treated with cyanamide without being isolated. Accurate control of the pH of the reaction mixture is required. The method can also be applied to hydroxylalkyl-, carbalkoxyalkyl- and carbalkoxy-α-aminocetals. A particularly advantageous way to prepare the intermediate α-alkylaminoaldehydes is by catalytic reduction of the α-alkylaminonitriles (XXIII) under controlled

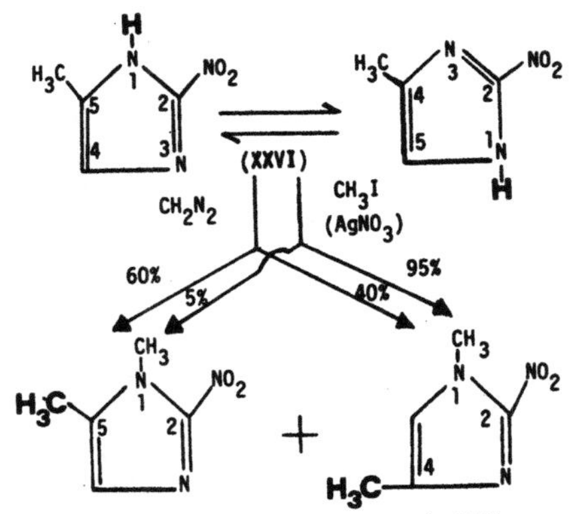

conditions (24,37). Akabori reduction of α-alkylaminoactones (XXIV) gives the α-alkylaminohydroxyalkylaldehydes (XXV), which are condensed as above to give 5-hydroxyalkyl-1-alkyl-2-aminoimidazoles (21,25). In some cases the 2-aminoimidazoles are isolated as picrates and then converted into the hydrochlorides.

The relatively simple 2-aminoimidazoles that were transformed into the corresponding 2-nitroimidazoles, as described in the previous section, were obtained by the methods outlined above.

Some other routes for synthesizing variously substituted 2-aminoimidazoles are also reported in the literature (38,39,40).

E. N-Substitution of 2-nitroimidazoles

The synthetic pathways described in Section C are almost the only ones that give 2-nitroimidazoles with substituents on both the 5-position of the nucleus and on the imine nitrogen. Methylation of 4(5)-methyl-2-nitroimidazole (XXVI) with methyl iodide in the presence of silver nitrate gives 1,4-dimethyl-2-nitroimidazole(XXVII) while with diazomethane, one obtains a 40/60 mixture of (XXVII) and

SCHEME VI

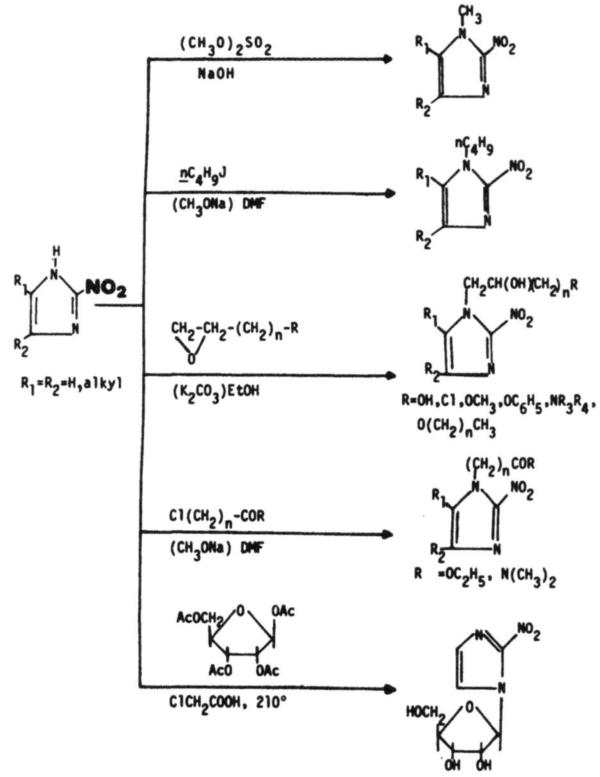

of 1,5-dimethyl-2-nitroimidazole (XXVIII) (4).

When there are no substituents in the 4(5)-position or when both positions have the same substituent, there is no problem of isomer formation. Therefore, a large number of 2-nitroimidazoles with substituents on the imine nitrogen have been reported in papers and in several patent applications (largely from Hoffmann-La Roche). They were prepared by reacting various alkylating agents (halides, epoxides, alkyl sulfates) (20,41,42,43) with sodium salts or in a basic medium. One of these 2-nitroimidazoles is Misonidazole* (Scheme VI). Azomycin-riboside has been prepared by fusion of a mixture of 2-nitroimidazole and tetra-O-acetyl-β-D-ribofuranose in the presence of a catalytic amount of chloroacetic acid (44).

F. 2-Nitroimidazole-5-carboxaldehydes

Although 2-nitroimidazoles whose 1 and/or 5 positions carry substituents that have functional groups have been transformed into

* α-(methoxymethyl)-2-nitro-1H-imidazole-1-ethanol.

SCHEME VII

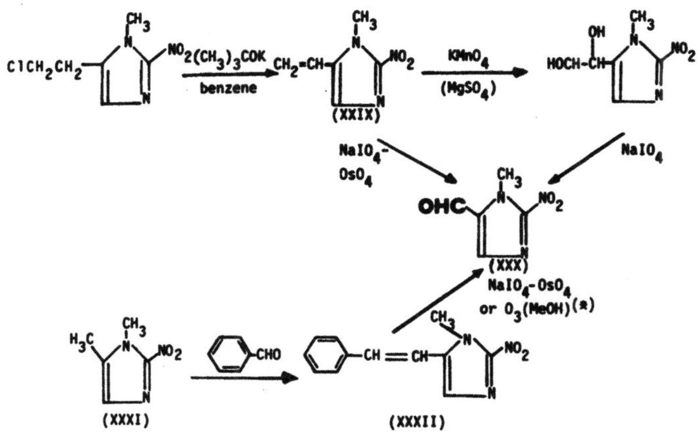

Cavalleri et al. (1972)
(*)Asato and Berkelhammer(1972)

many different compounds (21,22,23,41,42,45), it became possible to synthesize a large series of derivatives in position 5 only after 1-methyl-2-nitro-5-carboxaldehyde·(XXX) had been synthesized.

In fact, one of the reasons for the relatively minor development of the 1,5-disubstituted 2-nitroimidazoles is that each compound can be made only by first making the corresponding intermediate. It has been reported (4), and recently confirmed (46) that hydroxymethylation of 2-nitroimidazoles does not give the suitable intermediates for oxidation to the aldehydes. Several attempted synthetic routes, such as oxidation of 1,5-dimethyl-2-nitroimidazole (XXXI) with selenium dioxide, chromic anhydride or cerium ammonium nitrate were unsuccessful.

The compound was obtained (22,23) through the 5-vinyl intermediates (XXIX, XXXII), as shown in Scheme VII.

G. Derivatives in Position 5

In Scheme VII are shown some of the derivatives that can be synthesized from 1-methyl- and 1-ethyl-2-nitro-5-carboxaldehydes (22). Among these is a large series of hydrazones with various substituents (47,48,49), nitrones (49,50), Schiff's bases (48,51), oximes (48) and vinyl homologues (52). 5-Carboxaldehyde (XXX) is also an intermediate for preparing 2-nitroimidazoles bound through a carbon-carbon bond to another heterocyclic ring (XXXIII(23), XXXIV(52), XXXV(53)). 5-Hydroxymethyl-1-methyl-2-nitroimidazole (XXXVI) has been obtained by $NaBH_4$ reduction of (XXX)(45).

SCHEME VIII

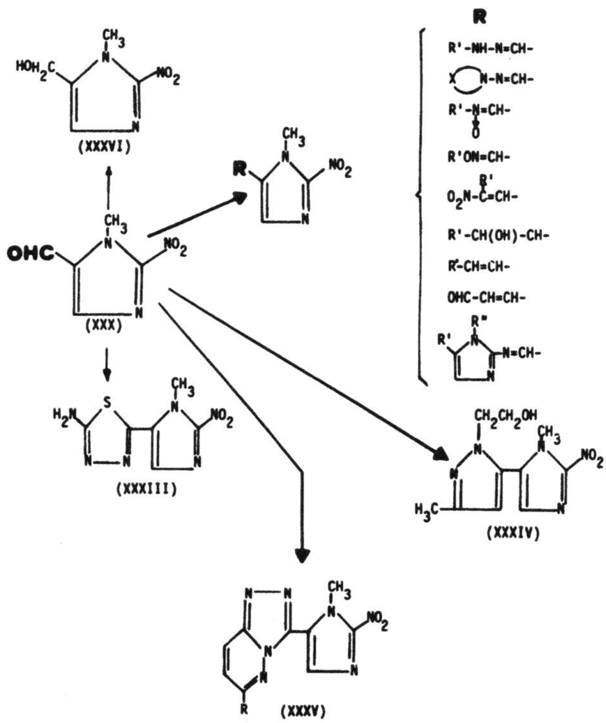

II. 5-NITROIMIDAZOLES

A. Introduction of the Nitro Group

Ever since the earliest work of Pyman (54), it has been known that nitration of imidazoles by conventional methods leads to introduction of a nitro group into the equivalent positions 4 or 5, if the imine nitrogen is not substituted.

By some variations aimed at controlling the reaction (8,55-58), various 4- and 5-nitroimidazoles carrying halogen, alkyl (8,56,59), aryl (57-60), imidazolyl (61,62), 1,3,4-thiadiazolyl (58) or nitro (9) substituents have been prepared. Some, especially those with 2-methyl or 2-isopropyl chains, have been widely used as intermediates.

When the imine nitrogen is substituted, isomers are formed and the nitro group may enter either position 4 or 5. In general, the 4-nitroimidazoles predominate (12). Selective nitrations have been reported. Treatment of 1-phenylimidazole (XXXVII) with

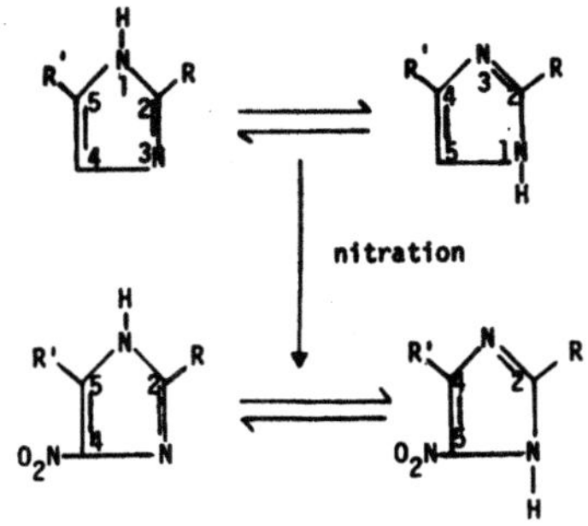

R = H, halogen, nitro, hydroxymethyl, acetoxymethyl, alkyl, aryl,imidazolyl, acetylamino-1,3,4-thiadiazolyl

R' = H, halogen, alkyl, aryl

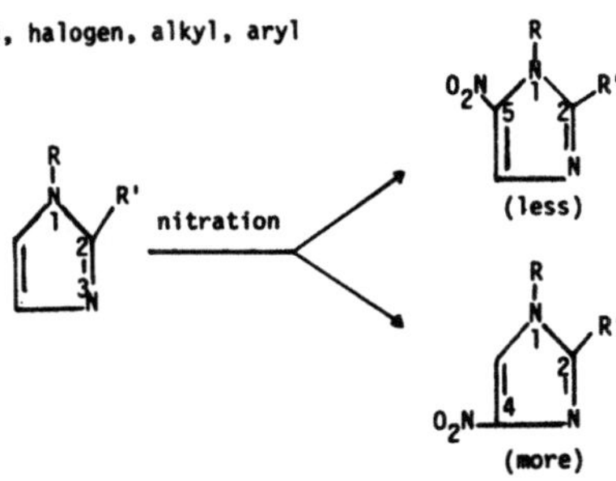

R = alkyl, aryl, imidazolyl, acetylamino-1,3,4-thiadiazolyl

R'= H, halogen, alkyl, aryl, sulfide

nitronium tetrafluoborate in chloroform gives the 5-nitro compound (63) (Scheme IX).

For the 2,2'-biimidazoles (XXXVIII), the orientation is determined by the presence of substituents and by the ratio of substrate to nitric acid (61,62).

A highly specific procedure for nitration at C-5 of 2-amino-5-(1-methyl-2-imidazolyl)-1,3,4-thiadiazole (XXXIX) consists of heating the compound with acetic anhydride and then adding acetic acid and a large excess of 70% nitric acid (58). When 1-alkyl-2-imidazolyl sulfides are treated at 100° with aqueous nitric acid, the 5-nitro derivative is formed preferentially (64).

A specific way to obtain 4-nitroimidazoles is by thermal decomposition of the quaternary iodides to give the thermodynamically more stable 4-isomer (65). This method has been applied (66) to the preparation of 1-methyl-2-isopropyl-4-nitroimidazole (XL) isomer free (Scheme IX; Eq. 10). 1-Methoxymethyl-2-(4-fluorophenyl)-4-nitroimidazole (XLI), an intermediate for the synthesis of

SCHEME IX

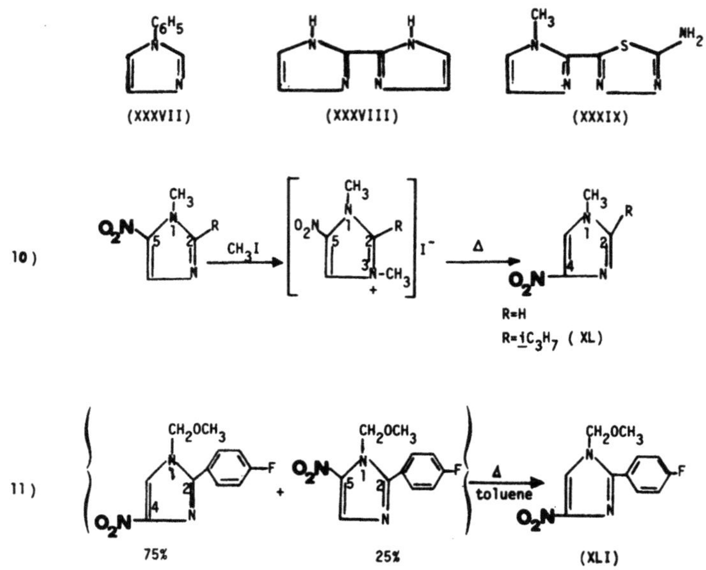

Flunidazole*, was obtained (67) by heating a mixture of the two isomers prepared by alkylation (Eq. 11).

An atom of iodine can be displaced by the nitro group at room temperature (68). This method has been used (69) to prepare 2-[1-methyl-1-hydroxy (or methoxy)-ethyl]-4(5)-nitroimidazole (XLII) (Scheme X).

B. N-Substitution of 4(5)-nitroimidazoles

The N-alkyl-substituted 5-nitroimidazoles cannot be made practically by nitration of the N-alkylimidazoles. However, the conditions for selective alkylation of 4(5)-nitroimidazoles to give 1-alkyl-4-nitro- or 1-alkyl-5-nitroimidazoles are quite well defined. The initial observations (65,70), that the reaction of methyl sulfate with 4(5)-nitroimidazole gave 1-methyl-4-nitro- or 1-methyl-5-nitroimidazole, depending on whether or not base was added, have been followed by investigation by many other researchers (71-74). The effects of pH of the medium on the site alkylated has been established (8,55,59,61,62,67,75,76): alkylation with alkyl halides or sulfates under alkaline conditions gives primarily the 4-nitro isomer, because in the anion, the nitrogen in position 3 is the more basic and therefore the more nucleophilic. Under neutral or weakly acidic conditions, the un-protonated nitrogen is alkylated, depending more on the tautomeric equilibrium. If an excess of the

* 2-(4-fluorophenyl)-5-nitro-1H-imidazole-1-ethanol.

SCHEME X

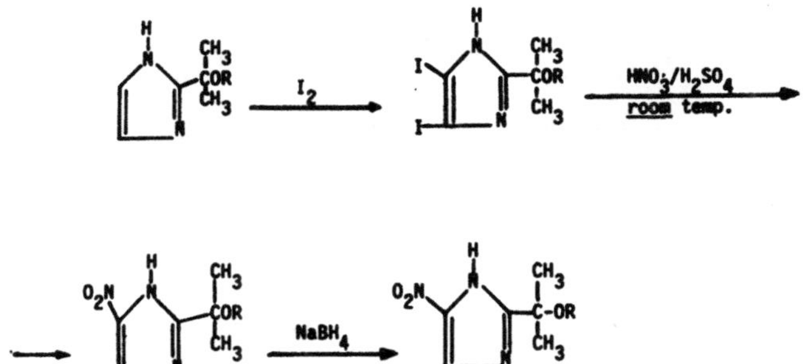

R = H, CH_3

(XLII)

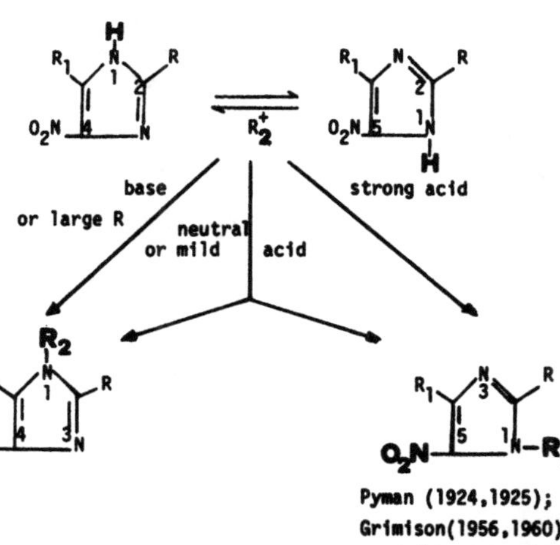

Pyman (1924,1925); Cosar (1966)
Grimison(1956,1960); Butler (1967)
Kaifež(1967,1968); Miller (1970)

TABLE III

N^1-SUBSTITUTION IN 4(5)-NITROIMIDAZOLES

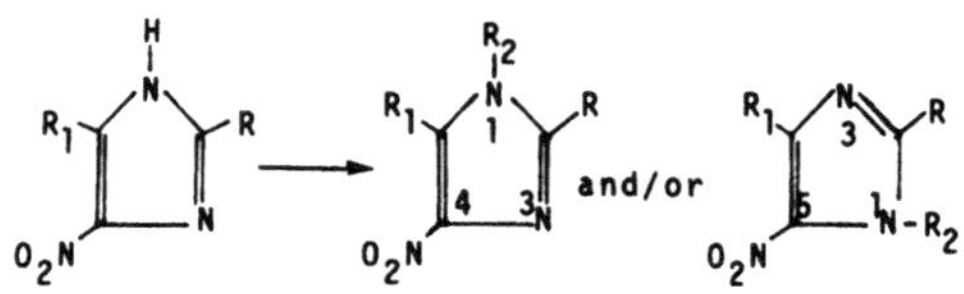

R = H, alkyl, aryl, NO_2, CH_2OH, I, Cl, imidazolyl

R_1 = H, aryl, styryl, SR_3, SO_2NH_2

Alkylating agents	R_2
$(CH_3O)_2SO_2$	CH_3
$ClCH_2OCH_3$	CH_2OCH_3
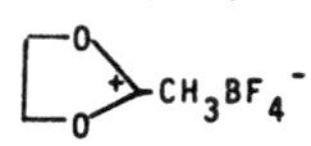	CH_2CH_2OAc
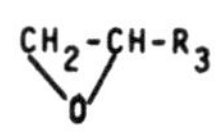	$CH_2-CH(OH)-R_3$
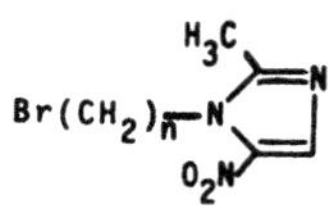	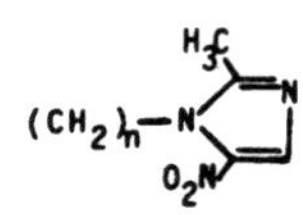
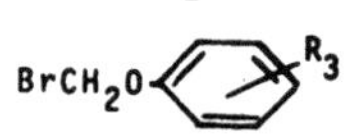	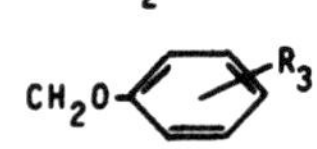
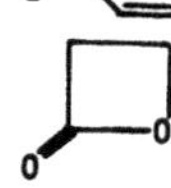	CH_2CH_2COOH

Alkylating agents	R_2
	$(CH_2)_3SO_3H$
$BrCH_2CH_2SC_2H_5$	$CH_2CH_2SC_2H_5$
$TsCH_2CH_2OH$	CH_2CH_2OH
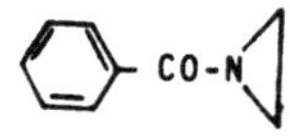	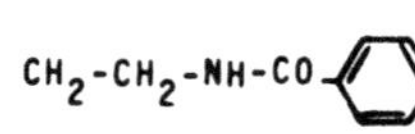
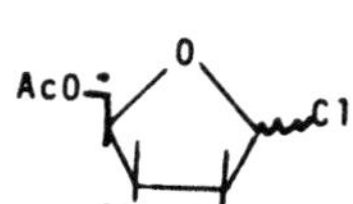	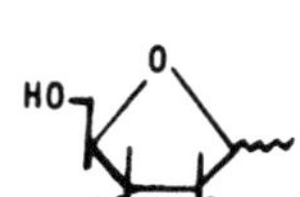
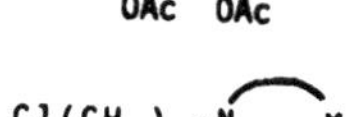	 X=O, CH_2
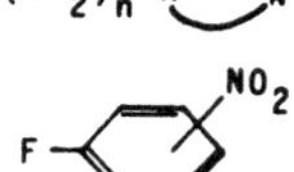	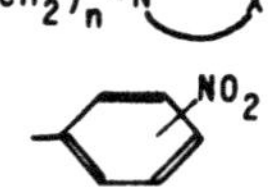

alkylating agent is added, the 4-nitro compound is again favoured. Finally, under strongly acidic conditions, the 5-nitro compound is formed preponderantly. There are some limitations to these "rules". Kaifez (75) has studied the effects of steric hindrance from the substituent groups on position 2.

4(5)-Nitroimidazoles with various types of substituents on carbon-2 or on the free carbon of the ring, or on both positions have been reacted (Table 3).

The various substituents include alkyl groups (8,55,59,75-78), styryls (79,80), aryls (60), imidazoles or nitroimidazoles (61,62), sulfonamide (81), nitro (9,10), iodo (68), alkoxy (68) and 2-amino-1,3,4-thiadiazole (58). Some examples of the alkylating agents that have been used are shown in the Table. They include alkyl (8) and aryl halides, sulfates and tosylates (8,59), lactones (55), 1,2-epoxides (59,77), dichloroalkanes (76), haloalkylacetates (76), haloalkylamines (78), haloalkylethers (67,76), onium salts (67), glycosides (82), haloalkylacetals (83), haloalkylsulfides (84), haloalkylimidazoles (85) and N-benzoylaziridine (86).

The substituents that have been introduced onto the imine nitrogen to give 5-nitro and 4-nitroimidazoles include hydroxy-, chloro-, bromo-, thio-, alkoxy-, carbalkoxy-, cyano-, amino-alkyls and -aryls, alkylsulfones, etc. Some of the compounds prepared are in therapeutic use, while others are intermediates with functional groups that make it possible to obtain a very large number of derivatives.

A survey of the physico-chemical properties of the nitroimidazoles has been made by Sunjić (87). The introduction of the nitro group reduces the basicity of the imidazole itself. The presence and the position of the nitro group and, when present, of other substituents on the ring nitrogen, cause differences that can be utilized both for separation of the isomers (basicity, solubility) and for analytical identification (pK_a, UV, IR, NMR, E1/2).

C. 5-Nitroimidazoles with Functional Groups in Position 2 (CHOH, CHO, CH_2Cl)

In 1964 Merck applied for a patent (88) for 1-methyl-5-nitro-2-carboxaldehyde (XLIV) opening the way for synthesis of 1-alkyl-5-nitroimidazoles with different substituents in position 2 (Scheme XI).

Several procedures (88,92) have been used for the preparation of (XLIV), but the one that gives 5-nitro-2-carboxaldehydes N-substituted with alkyls, benzyloxy- and acetoxy-alkyl, hydroxyethyl (after hydrolysis), or allyl groups, in fairly good yields, is oxidation of the corresponding 2-hydroxymethylimidazoles (XLIII) with manganese dioxide (90,91) or lead tetraacetate (92) (Eq. 14).

SCHEME XI

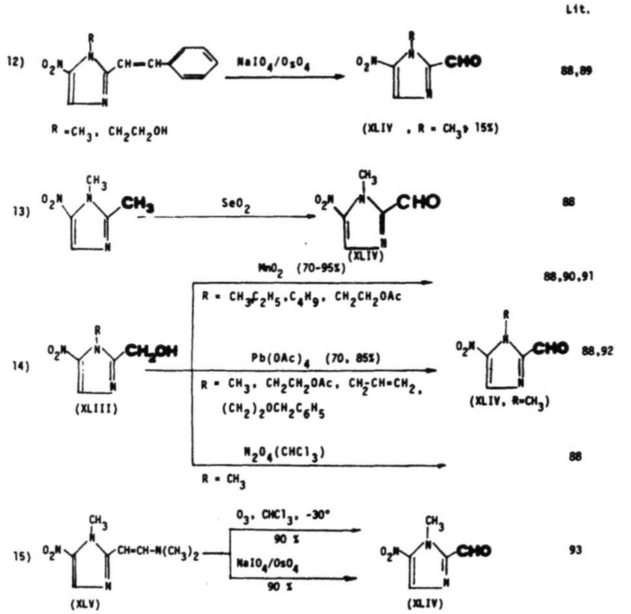

Rufer described (93) an interesting approach in which 2-(2-dimethyl-aminovinyl)-1-methyl-5-nitroimidazole (XLV), obtained by condensation of t-butoxymethylene-bis-dimethylamine with 1,2-dimethyl-5-nitro-imidazole, is oxidized with ozone or with osmium tetroxide-sodium periodate (Eq. 15). As mentioned above, 5-nitroimidazoles with a hydroxymethyl in position 2 are the most widely used precursors for the synthesis of 2-aldehydes. 1-Methyl-2-hydroxymethyl-5-nitro-imidazole (XLIII), described (94) for the first time in a patent in 1964, can be prepared by treatment of 1-methyl-5-nitroimidazole (XLVI) with paraformaldehyde in various solvents (DMF(95), DMSO (90,92), or chlorinated benzenes (96)).

Hydroxymethylation of 5-nitro-, 4(5)-nitro- and 2-nitro-imidazoles has been recently reinvestigated (46). While it does not go in the last two cases, it gives 60-70% yields when applied under the conditions outlined above to 5-nitroimidazoles with alkyl, hydroxy-, acetoxy-, chloro-, benzyloxyalkyl or allyl groups in position 1 (Scheme XII).

The 2-hydroxymethyl function present in variously 1-substituted-5-nitroimidazoles has been converted by conventional chemical procedures into 2-chloromethyl (98) and 2-carbonyl chloride (99), while the 2-carboxaldehyde has been transformed into the 2-carbonitrile directly by treatment with hydroxylamine-O-sulfonic acid (46) or

SCHEME XII

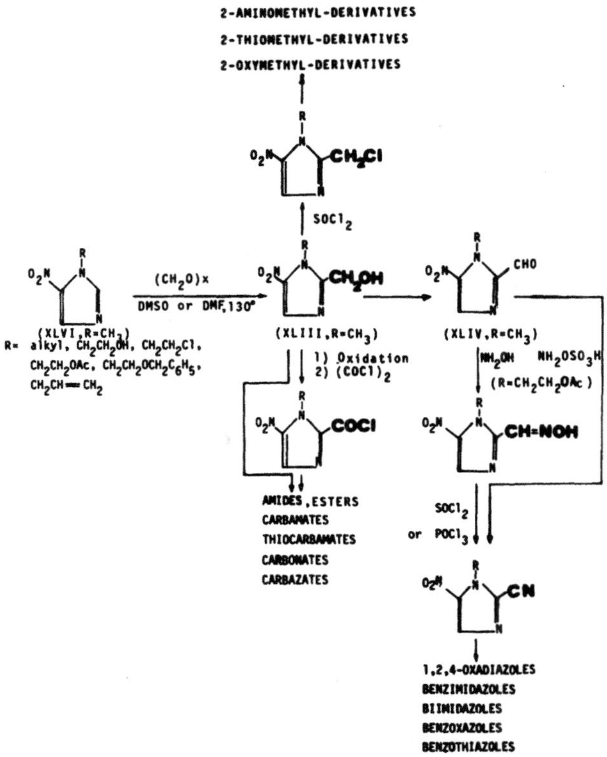

indirectly through the corresponding 2-carboxaldehyde-oxime (61, 100).

D. Derivatives in Position 2

The possibilities for synthetic transformations offered by the 2-substituents described above, and particularly by 2-carboxy-aldehyde, have been widely exploited (Scheme XII and XIII). Winkelmann has reviewed (101-104) a large number of derivatives, some of them previously described in patent applications.

From 2-hydroxymethyl-5-nitroimidazoles, by reaction with aryl intermediates having reactive halogen (fluorine or chlorine), a series of substituted 2-oxymethyl derivatives has been prepared (103). 2-Chloromethyl-5-nitroimidazoles have been reacted with hydroxy intermediates (alcohols, phenols) (102,103), with amines, heterocyclic bases or mercapto compounds (101-104). 2-(1,2,4-Oxadiazolyl)-(105), 2-imidazolyl-(61), and 2-benzimidazolyl-5-nitroimidazoles (100) have been obtained by 2-carbonitrile transformations. A few examples of derivatives obtained by the reaction

SCHEME XIII

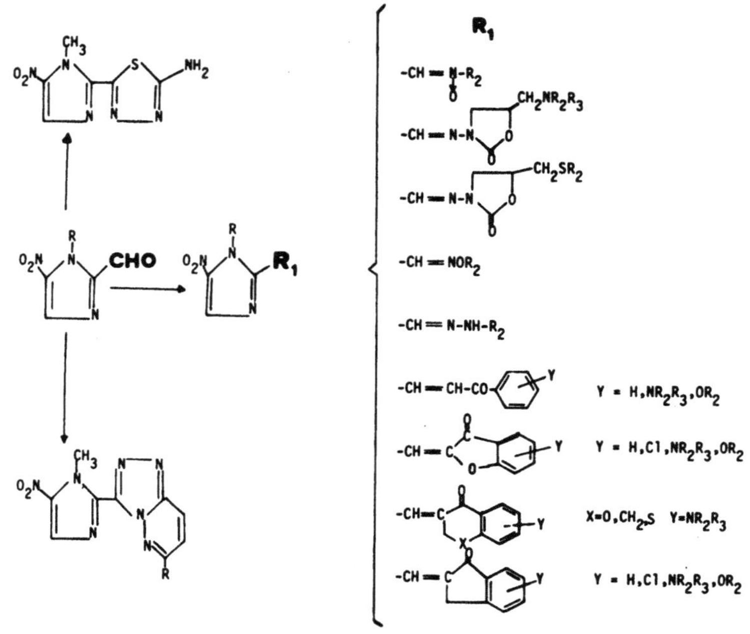

of N-substituted-5-nitroimidazole-2-carboxaldehydes with suitable reagents are summarized in Scheme XIII. The substituents are linked to the nitroimidazole nucleus through a methylene or directly through a carbon-carbon bond. Among these are N-substituted nitrones (91), oximes (101), hydrazones (101), aminomethyl- and alkylthiomethyl-oxazolidinones (90,106,107), indanones (92), amino-1,3,4-oxadiazoles and amino-1,3,4-thiadiazoles (108), pyrazoles (109), amino- and alkoxyamino-indanones, tetralones and acetophenones (110,111), tetralones, benzofuranones, chromanones and some thioanalogues (112), Δ^2-1,3,4-oxadiazolines (113) and thiazolopyridazines (114).

1. H. Horie, J. Antibiotics Ser. A. 9:168 (1956).
2. S. Nakamura, Pharm. Bull. 3:379 (1955).
3. K. Maeda, T. Osato, and H. Umezawa, J. Antibiotics, Ser. A. 6:182 (1953).
4. G. C. Lancini, N. Maggi, and P. Sensi, Il. Farmaco, Ed. Sci. 18:390 (1963).
5. F. Rung, and M. Behrend, Ann. 271:28 (1892).
6. A. Windaus, Ber. 42:758 (1909).
7. R. M. Jacob, G. L. Régnier, and C. Crisan, U.S. Patent 2, 944,061, July 1960 - Chem. Abstr. 55:1657 (1961).
8. C. Cosar, C. Crisan, R. Horclois, R. M. Jacob, S. Tchelitcheff, and R. Vaupré, Arzneim-Forsch. (Drug. Res.), 16:23 (1966).

9. K. C. Agrawal, K. B. Bears, and R. K. Sehgal, J. Med. Chem. 22:583 (1979).
10. R. K. Sehgal, and K. C. Agrawal, J. Heterocycl. Chem. 16:1499 (1979).
11. J. Suwiński, E. Salwinska, J. Watras, and M. Widel, Abstracts of the "Int. Conference on the Chemistry, Pharmacology and Clinical Application of Nitroimidazoles", Cesenatico (Italy) August 1980, page 46.
12. K. Hofmann, "Imidazole and Its Derivatives", Part I, in: "The Chemistry of Heterocyclic Compounds", Interscience Publishers Inc., New York (1953).
13. G. C. Lancini, D. Kluepfel, E. Lazzari, and G. Sartori, Biochim. Biophys. Acta, 130:37 (1966).
14. Y. Seki, T. Nakamura, and Y. Okami, Biochem. J. 67:389 (1970).
15. G. C. Lancini, E. Lazzari, and G. Sartori, J. Antibiotics 6:387 (1968).
16. A. F. Pozharskii, E. A. Zvezdina, and A. M. Simonov, Tetrahedron Letters, 23:2219 (1967).
17. M. D. Coburn, J. Heterocycl. Chem. 7:455 (1970).
18. G. C. Lancini, and E. Lazzari, Experientia 21:1 (1965).
19. A. G. Beaman, W. Tautz, T. Gabriel, and R. Duschinsky, J. Am. Chem. Soc. 87:389 (1965).
20. A. G. Beaman, W. Tautz, T. Gabriel, O. Keller, V. Toome, and T. Duschinsky, Antim. Agents Chemother. 1965:469
21. G. C. Lancini, E. Lazzari, V. Arioli, and P. Bellani, J. Med. Chem. 12:775 (1969).
22. B. Cavalleri, R. Ballotta, and G. C. Lancini, J. Heterocycl. Chem. 9:979 (1972).
23. G. Asato, and G. Berkelhammer, J. Med. Chem. 15:1086 (1972).
24. B. Cavalleri, G. Volpe, V. Arioli, and G. C. Lancini, J. Med. Chem. 20:1522 (1977).
25. B. Cavalleri, G. Volpe, V. Arioli, F. Pizzocheri and A. Diena, J. Med. Chem. 21:781 (1978).
26. G. Sartori, G. C. Lancini, and B. Cavalleri, J. Labelled Compounds and Radiopharmaceuticals, XV:673 (1978).
27. W. A. Sheppard, and O. W. Webster, J. Am. Chem. Soc. 95:2695 (1973).
28. O. W. Webster, Ger. Offen. 2,3174,53, Oct. 1973 - Chem. Abstr. 80:14926 (1974).
29. O. V. Lebedev, et al., U.S.S.R. Patent 177869, Jan. 1966 - Chem. Abstr. 64:19630 (1966).
30. O. V. Lebedev, et al., U.S.S.R. Patent 184868, July 1966 - Chem. Abstr. 67:90809 (1967).
31. B. A. Tertov, V. V. Burykin and A.S. Morkovnik, U.S.S.R. Patent 437,763, July 1974 - Chem. Abstr. 81:169542 (1974).
32. J. Martin, and F. Johnson, U.S. Patent 3,828,064, Aug. 1974 - Chem. Abstr. 81:120625 (1974).
33. G. C. Cherkofsky - Europ. Patent Appl. 0,008,775 - March 1980.
34. D. P. Davis, K. L. Kirk, and L. A. Cohen, Abstracts of the "International Conference on the Chemistry, Pharmacology

and Clinical Application of Nitroimidazoles", Cesenatico (Italy) August 1980, page 52.
35. A. Lawson, J. Chem. Soc.1955:1695.
36. G. C. Lancini, and E. Lazzari, J. Heterocycl. Chem. 3:152 (1966).
37. M. Stagi, and A. Gianantonio, Ger. Offen. 2,600,537, July 1976 - Chem. Abstr. 85:160097 (1976).
38. W. Nagai, K. L. Kirck, and L. A. Cohen, J. Org. Chem. 38: 1971 (1973).
39. T. Nishimura, and K. Kitajima, J. Org. Chem. 41:1590 (1976) ibid. 44:818 (1979).
40. M. Ruccia, N. Vivona, and G. Cusmano, Tetrahedron, 30:3859 (1974).
41. A. G. Beaman, W. Tautz, and R. Duschinsky, Antim. Agents Chemother. 1967:520.
42. W. Tautz, S. Teitel, and A. Brossi, J. Med. Chem. 16:705 (1973).
43. M. Hoffer, and E. Grunberg J. Med. Chem. 17:1019 (1974).
44. R. J. Rousseau, R. K. Robins, and L. B. Towsend, J. Heterocycl. Chem. 4:311 (1967).
45. B. Cavalleri, R. Ballotta, V. Arioli, and G. C. Lancini, J. Med. Chem. 16:557 (1973).
46. R. K. Sehgal, and K. C. Agrawal, J. Heterocycl. Chem. 16:871 (1979).
47. B. Cavalleri, G. Volpe, and R. Pallanza, Arzneim.-Forsch. (Drug Res.) 25:148 (1975).
48. B. Cavalleri, G. Volpe, A. Ripamonti, and V. Arioli, Arzneim.-Forsch. (Drug Res.) 27:1131 (1977).
49. B. Cavalleri, R. Ballotta, and V. Arioli, Arzneim.-Forsch. (Drug Res.) 25:338 (1975).
50. B. Cavalleri, G. Volpe, V. Arioli, and G. C. Lancini, Arzneim.-Forsch. (Drug Res.) 27:1391 (1977).
51. A. Malabarba, M. Berti, A. De Paoli, and B. Cavalleri, Il Farmaco, Ed. Sci. 34:105 (1979).
52. B. Cavalleri, G. Volpe, V. Arioli, J. Med. Chem. 20:656 (1977).
53. H. Berger, R. Gall, K. Stach, W. Voemel, and R. Hoffmann, Ger. Offen. 2,261,693, June 1974 - Chem. Abstr. 81:120674 (1974).
54. R. G. Fargher and F. L. Pyman, J. Chem. Soc. 115:217 (1919) ibid. 115:234 (1919).
55. M. W. Miller, H. L. Howes Jr., R. V. Kasubick, and A. R. English J. Med. Chem. 13:849 (1970).
56. A. Zmojdzin et al. Pol. Patent 92,065, Dec. 1977 - Chem. Abstr. 90:87460 (1979).
57. J. S. Amato, V.J. Grenda, T. M. H. Liu, and E. J. J. Grabowski, J. Heterocycl. Chem. 16:1153 (1979).
58. W. A. Remers, G. J. Gibs, and M. J. Weiss, J. Heterocycl. Chem. 6:835 (1969).
59. K. Butler, H. L. Howes, J. E. Lynch, and D. K. Pirie, J. Med. Chem. 10:891 (1967).
60. G. P. Ellis, C. Epstein, C. Fitzmaurice, L. Goldberg, and G. H. Lord J. Pharm. Pharmacol. 16:801 (1966), ibid 19:102 (1967).

61. P. Melloni, E. Dradi, W. Logemann, I. de Carneri, and F. Trane, J. Med. Chem. 15:926 (1972).
62. P. Melloni, R. Metelli, De. Fusar Bassini, C. Confalonieri, W. Logemann, I. de Carneri, and F. Trane, Arzneim.-Forsch. (Drug Res.) 25:9 (1975).
63. Merck and Co. Inc., Neth. Appl. 66,09553, Jan. 1967 - Chem. Abstr. 67:11487 (1967).
64. R. C. Tweit, E. M. Kreider, and R. D. Muir, J. Med. Chem. 16:1161 (1973).
65. C. E. Hazeldine, F. L. Pyman, and J. Winchester, J. Chem. Soc. 1924: 1431.
66. M. Hoffer, M. Mitrovic, A. Beaman, and A. Brossi, J. Med. Chem. 14:993 (1971).
67. E. J. J. Grabowski, T. M. H. Liu, L. Salce, and E. F. Schoenewaldt, J. Med. Chem. 17:547 (1974).
68. M. Hoffer, V. Toome, and A. Brossi, J. Heterocycl. Chem. 3:454 (1966).
69. M. Hoffer, and A. MacDonald Jr., U.S. Patent 3,652,579, March 1972 - Chem. Abstr. 76:153741 (1972).
70. W. G. Forsyth, and F. L. Pyman, J. Chem. Soc. 1925: 573.
71. A. Grimison, and B. V. Smith, Chem. Ind. (London) 1956: 983.
72. A. Grimison, J. H. Ridd and B. V. Smith, J. Chem. Soc. 1960: 1352; ibid. 1960: 1357.
73. J. H. Ridd, and B. V. Smith, J. Chem. Soc. 1960:1363.
74. E. Laviron, Bull. Soc. Chim. France 1963: 2840.
75. F. Kajfež, D. Kolbah, M. Oklobdžija, T. Fajdiga, M. Slamnik, and V. Šunjić, Croat. Chem. Acta. 39:199 (1967)
76. F. Kajfež, V. Šunjić, D. Kolbah, T. Fajdiga, and H. Oklobdžija, J. Med. Chem. 11:167 (1968).
77. V. Šunjić, D. Kolbah, F. Kajfež, and N. Blažević J. Med. Chem. 11:1264 (1968).
78. P. N. Giraldi, V. Mariotti, G. Nannini, G. P. Tosolini, E. Dradi, W. Logemann, I. de Carneri, and G. Monti Arzneim.-Forsch. (Drug Res.) 20:52 (1970).
79. P. N. Giraldi, V. Mariotti, and I. de Carneri, J. Med. Chem. 11:66 (1968).
80. W. J. Ross, and A. Todd J. Med. Chem. 16:863 (1973).
81. B. S. Huang, J. M. Lauzon, and J. C. Parham, J. Heterocycl. Chem. 16:811 (1979).
82. C. Chavis, F. Grodenic, and J. L. Imbach, Eur. J. Med. Chem. 14:123 (1979).
83. S. S. Berg, and B. W. Sharp, Eur. J. Med. Chem. 10:171 (1975).
84. V. Kaplar, V. Šunjić, and F. Kajfež J. Heterocycl. Chem. 11:1055 (1974).
85. V. Šunjić, T. Fajdiga, N. Blažević, and F. Kaifež Acta Pharm. Jug. 19:65 (1969).
86. J. Heeres, J. H. Mostmans, B. Maes, and L. J. J. Backx, Eur. J. Med. Chem. 11:237 (1976).
87. V. Šunjić, F. Kaijfež, D. Kolbah, and N. Blažević Die Pharmazie 27:131 (1972), and literature cited.

88. D. W. Henry, and R. D. Hoff, Belg. Patent 661,262, Sept. 1965 - Chem. Abstr. 64:2093 (1965).
89. G. Asato, and J. English, U.S. Patent 3,565,892, Feb. 1979 - Chem. Abstr. 75:36041 (1971).
90. C. Rufer, H.-J. Kessler, and E. Schröder J. Med. Chem. 14: 94 (1971).
91. R. E. Bambury, C. M. Lutz, L. F. Miller, H. K. Kim, and H. W. Ritter, J. Med. Chem. 16:566 (1973).
92. C. Rufer, and H.-J. Kessler, Chim. Ther. 7:122 (1972).
93. C. Rufer, and K. Schwarz Liebigs Ann. Chem. 1975:1465.
94. May and Baker Ltd., Belg. Patent 639,469, April 1964 - Chem. Abstr. 62:9142 (1965).
95. E. Winkelmann, and H. Kroha Ger. Offen. 2,425,292, Dec. 1975 - Chem. Abstr. 85:123919 (1976).
96. Merck and Co. Inc., Neth. Appl. 64,13814, May 1965 - Chem. Abstr. 63:18097 (1965).
97. J. M. Chemerda, J. Kollonitsch, and S. Marburg, U.S. Patent 3,548,007, June 1971 - Chem. Abstr. 75:49081 (1971).
98. Merck and Co. Inc., Neth. Appl. 65,03901, Sept. 1965 - Chem. Abstr. 64:12683 (1966).
99. Merck and Co. Inc., Neth. Appl. 64,09117, Feb. 1965 - Chem. Abstr. 63:606 (1965).
100. C. Rufer, H.-J. Kessler, and E. Schröder Chim. Ther. 6:286 (1971).
101. E. Winkelmann, W. Raether, U. Gebert, and A. Sinharay, Arzneim-Forsch. (Drug Res.) 27:2251 (1977).
102. E. Winkelmann, W. Raether, and A. Sinharay, Arzneim.-Forsch. (Drug Res.) 28:351 (1978).
103. E. Winkelmann, and W. Raether Arzneim.-Forsch. (Drug Res.) 28:739 (1978).
104. E. Winkelmann, W. Raether, and U. Gebert, Arzneim.-Forsch. (Drug Res.) 28:1682 (1978).
105. G. Asato, and G. Berkelhammer, U.S. Patent 3,795,735 March 1974 - Chem. Abstr. 80:120959 (1974).
106. H.-J. Kessler, C. Rufer, and K. Schwarz, Eur. J. Med. Chem. 11:19 (1976).
107. C. Rufer, H.-J. Kessler, and E. Schröder J. Med. Chem. 14:934 (1971).
108. G. Berkelhammer, and G. Asato, U.S. Patent 3,740,434, June 1973- Chem. Abstr. 79:66365 (1973).
109. J. P. Verge, M. C. Neville, and H. Friedman, Ger. Offen. 2,431, 775, Feb. 1975 - Chem. Abstr. 83:10079 (1975).
110. C. Rufer, H.-J. Kessler, E. Schröder, and A. Damerius, Chim. Ther. 8:567 (1973).
111. C. Rufer, H.-J. Kessler, and E. Schröder, J. Med. Chem. 18:253 (1975).
112. C. Rufer, H.-J. Kessler, E. Schröder, and A. Damerius, Chim. Ther. 7:5 (1972).
113. M. T. Wu, J. Heterocycl. Chem. 9:31 (1972).

114. H. Berger, R. Gall, and K. Stach, Ger. Offen, 2,261,735, June 1974 - Chem. Abstr. 81:120673 (1974).

CHEMICAL PROPERTIES AND REACTION MECHANISMS OF NITROIMIDAZOLES

A. Breccia

University of Bologna

Italy

INTRODUCTION

The chemical and physico-chemical properties of nitroimidazoles can be considered in terms of two principal aspects :

i) properties that can be correlated with the molecular structure, these are useful in guiding the synthesis, separation and identification of new derivatives;

ii) properties that influence their pharmacological behaviour.

The first aspect has been discussed by Dr. Cavalleri; the second forms part of the subject matter of this paper.

The reaction mechanisms of nitroimidazoles which will be discussed will concern therefore, the reactivities of these compounds with chemical species involved in the biological systems in which the nitroimidazoles can function as drugs.

The general molecular structures of nitroimidazoles from which all other derive, are 2-, 4-, and 5-nitroimidazoles. The first natural nitro-compound containing imidazole studied pharmacologically was the antibacterial agent, azomycin (2-nitroimidazole) (1). This was followed by metronidazole (2) (Flagyl) a 5-nitroimidazole derivative which, in many respects, is still the best available drug for the treatment of many anaerobic bacterial diseases.

In 1974 a new 2-nitroimidazole derivative, misonidazole, was first investigated clinically as a hypoxic cell sensitizer for use in radiotherapy. The structures of these drugs are shown below.

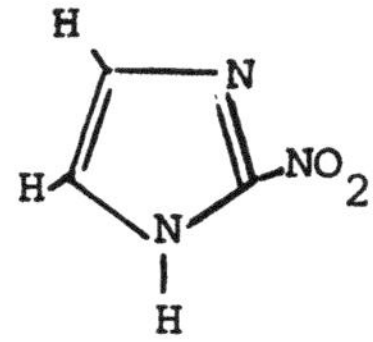

Azomycin

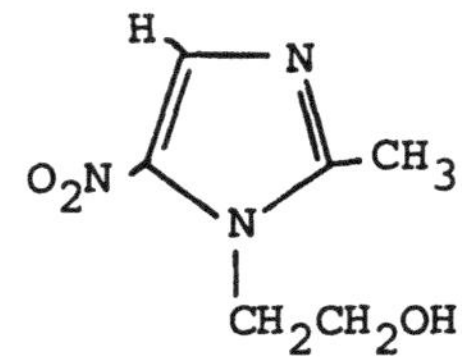

Metronidazole

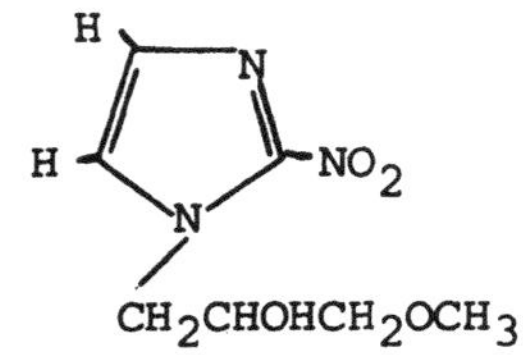

Misonidazole

Subsequently, many other derivatives have been prepared containing 2-, 4- and 5-nitro and dinitro groups. A few of these are already in use in pharmacology and others are under investigation as is discussed in the paper by Biagi.

RELATIONSHIP BETWEEN CHEMICAL, PHYSICO-CHEMICAL AND PHARMACOLOGICAL PROPERTIES OF NITROIMIDAZOLES

The nitroiminidazoles are widely used in different fields of pharmacological interest. They include : anaerobic infections such as tricomoniasis, amoebiasis and other diseases arising from obligate anaerobes and protozoa; hypoxic cell radiosensitizers in radiotherapy and possibly as adjuncts with various other anti-cancer drugs. Other applications have been suggested in some liver diseases and in arteriosclerosis where it is believed hypoxic cell may be present.

The pharmacological activity of the nitroimidazoles is confined mainly to cells or tissues where oxygen is absent or present only at a very low level. From the view point of studying chemical mechanisms, it is necessary therefore to use model systems under anaerobic conditions.

The main structural characteristics for the biological activity of these molecules appear to involve the presence of one or more nitro-groups giving the molecules a high (less negative) one-electron reduction potential. In some systems the steric protection of the NO_2 group by a side chain is necessary in order to reduce the rate of some metabolic processes and to influence acid/base properties.

It is also necessary to have information available on the physico-chemical properties of the biomolecules with which the nitroimidazoles react in biological systems. The majority of living organism or tissues require rather strict physico-chemical conditions although there are exceptions. In table 1, the minimum and maximum values or some relevant factors are shown (1).

Table 1. Range of environmental conditions

Factor	Lower limit	Upper limit
Temperature	$-18^{o}C$ (fungi, bacteria)	+ $104^{o}C$ (sulphate reducing bacteria under 1000 atm hydrostatic pressure
Redox potential	-450 mV at pH 9.5 (sulphate reducing bacteria)	+ 850 mV at pH 3 (iron bacteria)
pH	0 (Acontium velatum, fungus D, Thiobacilus thiooxidans)	13 (Plectonema nostocorum)
Hydrostatic pressure	Essentially zero	1400 atm (deep-sea bacteria)
Salinity	Double distilled water (heterothropic bacteria)	Saturated brines (Dunaliella, halophyllic bacteria, etc.)
Water activity	0.65-0.70 (Aspergillus glaucus)	Essentially 1

Information is available for some biological systems where nitroimidazoles have a demonstrable function. For example the redox potentials of some contents of the intestinal tract, necrotic tissues and abscess cavities are in the range -150 to -250 mV (2). This indicates the presence of hypoxia and in some cases, the local oxygen concentration can be as low as 10 ppm. Generally, the extra-cellular pH is lower as would be expected in oxygen-deficient regions. Cellular membrane potentials can be important also in influencing drug transport. For example the membrane potential of the organism Trichomonas Vaginalis is in the range 3-8 mV. Electrochemical properties are clearly particularly important in the interaction between nitroimidazoles and biological systems. Model systems suitable for studying these processes include electrochemical and radiation chemical methods where electron reactivity and electron transfer processes can often be measured directly.

Stradins and co-workers (3-10) carried out systematic investigations of the cathodic reduction of 2-nitrofuran derivatives at a dropping mercury electrode (DME) using classical and oscillographic polarography, cyclic voltammetry and other physico-chemical methods. The studies were aimed at investigating the rôle of redox potential in determining antibacterial activity. Later Stradins and co-workers (11) found that reduction of the NO_2 group or its elimination, led to the loss of biological activity. They extended their conclusions to other nitroaromatic derivatives and suggested two reaction mechanisms which may be partially applicable. Their conclusions were as follows:

a) The primary electrochemical process in aqueous solutions of materials of biochemical interest consists of the 4 e reduction of the nitro group to the hydroxylamine group. At more negative potentials, a further electroreduction to the amino group is possible with the reductive cleavage of the azomethine bond. However the last process takes place only at considerably higher cathodic potentials and are not of any direct interest to the biochemist.

b) The 4-electron reduction process consists of successive stages which vary in character according to the structures of the derivative and the solvent. The overall electrochemical process is irreversible although the first electron is transferred reversibly. Initially, a nitrofuran-anion-radical is formed, whose potential determines that of the total oxidation-reduction process. Transfer of the second electron proceeds irreversibly and is the cause of the irreversibility of the whole electrochemical process.

c) Nitrofuran molecules are strongly adsorbed at the charged electrode surface at the electro-reduction potential of the nitro group. Their electro-reduction is facilitated by superficial protonation of the nitro group and proceeds in the adsorbed state. Addition of a large quantity of an organic solvent or of a surface-active substance may lead to the partial desorption of nitrofurans. In turn, this inhibits electrochemical processes following the transfer of the first electron and thus protects the primary particles (free radicals and radical ions) from immediate conversion.

d) Suppression of the superficial protonation of nitrofuran molecules or of the radical-ions initially formed (pKa $\sim$ 5) allows a gradual change from the four electro-reduction process to occur which establishes conditions for the accumulation of free radicals.

A mechanism was proposed for the interaction of nitrofuran with the NAD and FAD redox enzyme systems which have redox potentials of the same order as that of the nitrofuran, see Plate A.

A second mechanism proposed concerned the chemistry of the polarographic reduction of 5-nitrofuran and its derivatives involving the formation of RNO_2^-. This is shown in the scheme below.

Inhibition of redox enzymes, FAD and NAD by nitrofuran

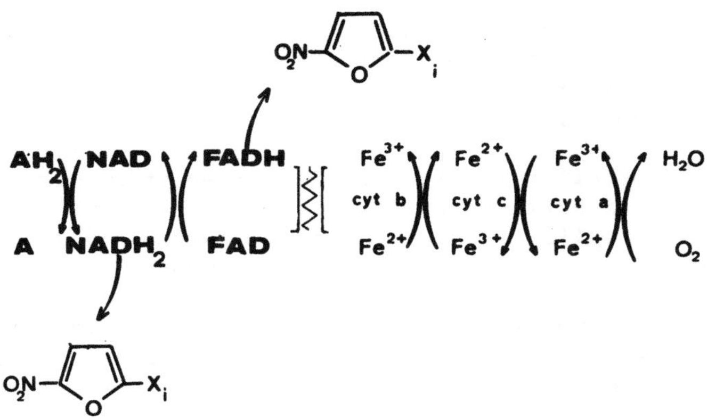

Mechanism of the polarographic reduction of 5-nitrofuran and its simplest derivatives

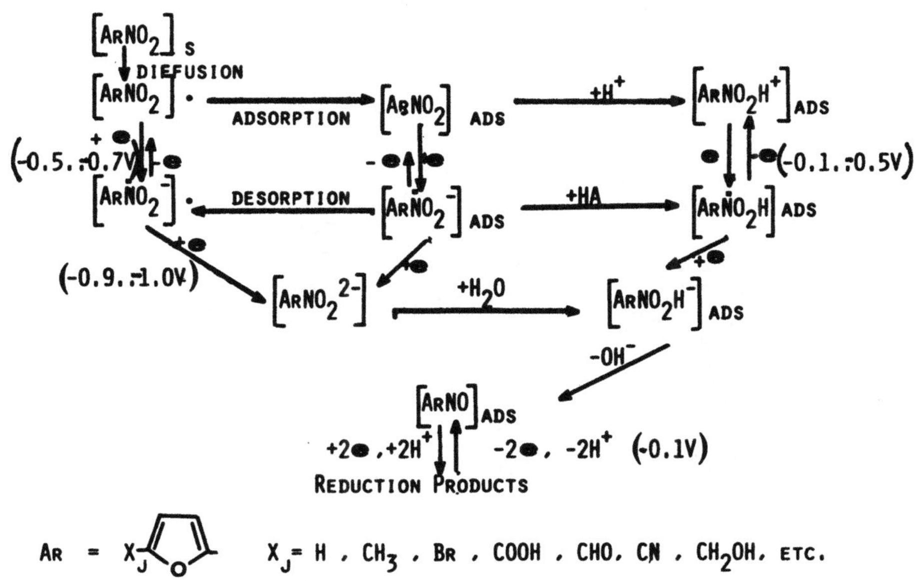

In 1963 Adams and Dewey suggested that a relationship existed between the ability of some chemical compounds to radiosensitize hypoxic cells and the electron affinities of these compounds (12). A correlation has been demonstrated experimentally (Fig.1) between the monoelectron reduction potentials, $E_7^{'}$, and sensitizing efficiency in irradiated mammalian cells (13) and in Fig.2 for E'_{pc} measured in an aprotic system (14).

The basis of all these mechanisms for anaerobic antibacterial and sensitizing effects is the occurrence of electron transfer processes in the cells. Chapman has discussed interaction with radiosensitizers according to the scheme shown below (15).

The first intermediate in the reaction mechanism for reduction of nitroaromatics, in particular nitroimidazoles, is the anion RNO_2^-. The electron which reacts with these compounds derives from two sources : one from biochemical systems, the other from radiation effects.

In the mechanism of the cytotoxic action in anaerobic bacteria only cellular electron donors are involved. In radiosensitization, both sources provide electrons to the system. There should be therefore two different mechanisms for the formation of RNO_2^- anions with probably different rate constants, i.e. the interaction of nitro-derivatives with redox biological systems and the reaction of radiation-reduced electrons or radicals with the compound.

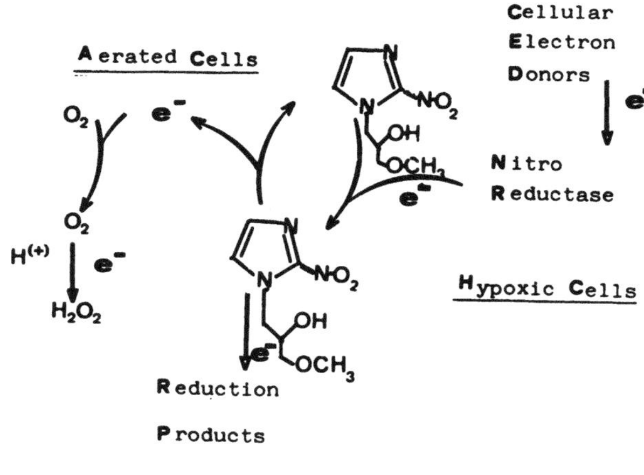

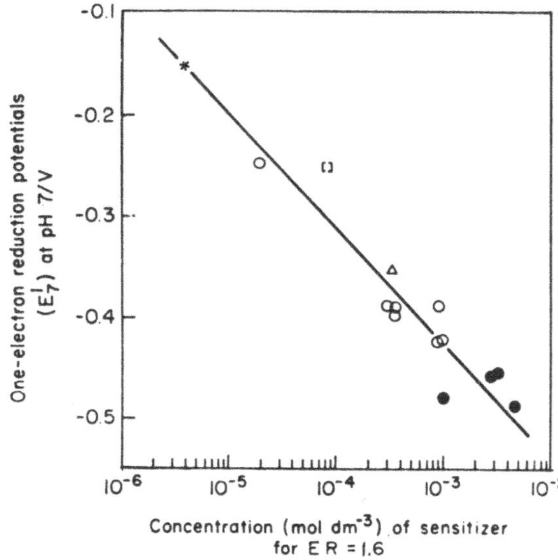

Fig.1 Dependence of radiosensitization efficiency in V 79-379A cells on one-electron reduction potential at pH = 7 for oxygen and other sensitizers.

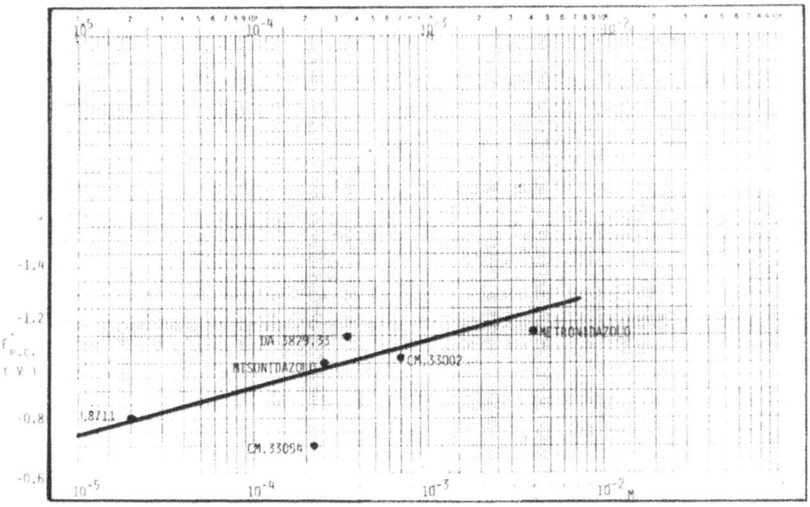

Fig.2 Peak potentials for the first cathodic process in DMF solution plotted against the sensitizer concentration for E.R. = 1.6. All compounds give reversible reactions with electrons except L 8711.

The rate constants of the last reactions are very fast as can be seen in Table 2 (16).

Table 2. Rate constants for reactions of e^-_{aq}

Compounds	Rate constants for reactions with e^-_{aq}
Compounds	k (M^{-1} s^{-1})
p. nitrotoluene	1.9×10^{10}
Nitrobenzene	3.0×10^{10}
Picric Acid	3.9×10^{10}
O_2	2.0×10^{10}
DNA	- 10^{12}
Metronidazole	2.2×10^{10}
DA 3827	1.8×10^{10}
Imidazole	3.7×10^{7}
DA 3838	2.6×10^{10}
Piridine	1.0×10^{9}
Adenosine	3.0×10^{10}

The speed of these reactions requires the use of fast-response techniques such as pulse radiolysis. The steps which occur after the formation of the radical-anion RNO_2^- can be different. One route involves the formation of the nitrite ion, NO_2^-, and the imidazolyl radical. In γ-radiolysis this accounts for 40-100% of the total transformation as is shown in Table 3.

Table 3.

Compounds		Dose krads	C_o μ M/ml		C_f μ M/ml		$C_{NO_2^-}$ μ M/ml		$C_{NO_2^-}$ / Compound transformed	
Metronidazole	160	50	4.67	0,1	4.29	0,060	0.83	0,021	100 %	56%
	320		4.67		3.97		0.66		95 "	
Misonidazole	160	50	1.00	0,1	0.60	0,069	0.15	0,013	40 "	42%
	320		1.00		0.60		0.23		40 "	
CM 54020 (5,nitro..)	160	50	0.49	0,1	0.25	0,051	0.22	0,045	90 "	90%
	320		0.49		0.15		0.24		70 "	
Desmethylmiso.		50		0,10		0,034		0,014		40%
L 87 11		50		0,042		0,025		0,0104		71%

Another route involves the further reduction of RNO_2^- to nitroso, the hydroxylamine and probably the amine. For metronidazole, but not for most other nitroimidazoles, the nitrite ion is formed during electrochemical reduction in a yield of 10% of the total yield of reduced compound (17). It is likely that the mechanism of formation of NO_2^- from metronidazole follows the scheme suggested below in which NO_2^- is formed as a result of oxidation by the OH radical (18).

1) 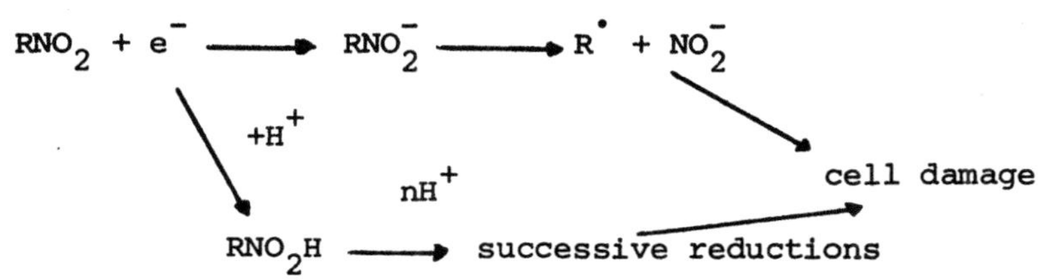

R = imidazolyl group

2) $RNO_2 + OH^{\bullet} \longrightarrow R(OH)NO_2^{\bullet} \longrightarrow H^+ + NO_2^- + R = O$

Electrochemical methods are very useful for the study of the interaction of nitroimidazoles with other biological molecules e.g. DNA (see Edwards this volume). A summary of the electrochemical methods available for the study of reaction mechanisms of nitroimidazoles is shown below.

Polarography	:	$E\frac{1}{2}$; E' with surface active substances (tylose)
Cyclic voltammetry	:	Electron reaction reversibility;
	:	$E\frac{1}{2}$ in aprotic systems;
	:	Reaction mechanisms of catalytic processes
Coulometry	:	n_{app} (number of electrons per mole of reduced compound)
Controlled potential electrolysis	:	n_{app} for large quantities;
	:	reduced derivatives production.

With respect to coulometry Table 4 shows results of calculations of the number of electrons involved in the reduction (19). Four electrons should be involved in the reduction of metronidazole to the corresponding hydroxylamine, but on the bases of n_{app} only three are involved. This means that no hydroxylamine is formed as was confirmed by chemical analysis.

Table 4. Electrochemical effects of sensitizers on cells (Euglena g)

Compounds	Cell surviving (Euglena) after electrolysis	E for electrolysis	n_{app} with cells	n_{app} without cells
Control	9.3 x 10	-1 V	-	-
Metronidazole	4.2 x 10	-0.7 V	4.2	3.0/Mole
Misonidazole	0.9 x 10	-0.7 V	4.3	3.8/Mole
L 87 11	0.6 x 10	-0.5 V	4.2	3.8/Mole
DA 38 29	2.5 x 10	-0.7 V	4.5	3.9/Mole
DA 38 38	0.9 x 10	-0.7 V	4.6	3.8/Mole
CM 54020	-1.4 x 10	-0.7 V	4.4	·3.9/Mole

However the reduction of the compounds in the presence of cells appears to involve more than four electrons. The overall process leads to cell killing, indicating that the toxic product produced in cellular reactions involving the radiosensitizer is a reduced compound.

The pulse radiolysis method (20) is a good tool for the study of many fast reactions and processes such as the measurement of one-electron potentials, the calculation of rate constants of electron reactions or the kinetics of transient species. The technique gives information on the molecular structure of short-lived intermediates such as free radicals or radical-ions. It is not limited by the time required for delivery of the radiation which produces many transient species in the irradiated system. Pulse radiolysis data are valuable therefore for studying primary radiation sensitization effects, for example when the mechanism of action is still at an early chemical stage (21).

Lipophilicity and acid/base equilibrium

Lancini and colleagues suggested that the regions of the 2- and 5-nitroimidazole structures involving C_2, C_5 and N_1, are important in the interaction of these molecules with molecular receptors (22). Blazevic and co-workers (23) proposed that the $-NO_2$ group in the 2 or 5 position with a side chain at the other position are the lipophilic sites. This is shown in Fig.3.

The lipophilicity of a substance is expressed by the partition coefficient P, which is the ratio of the equilibrium concentrations of the compound in octanol (lipid phase) and

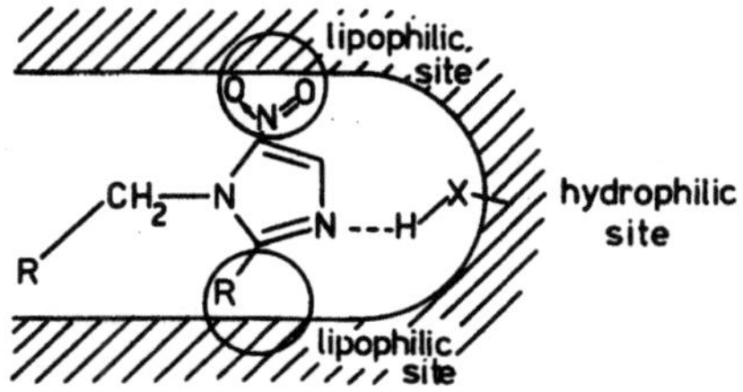

Fig. 3. Site of interaction of 2- and 5-nitroimidazoles with cellular receptors.

aqueous phase at a pH of 7.4. The higher the value of P, the more lipophilic is the compound. Data of Wardman (23) are shown in Fig.4, in which the values of log P are plotted against the Hantsch π substituent constant.

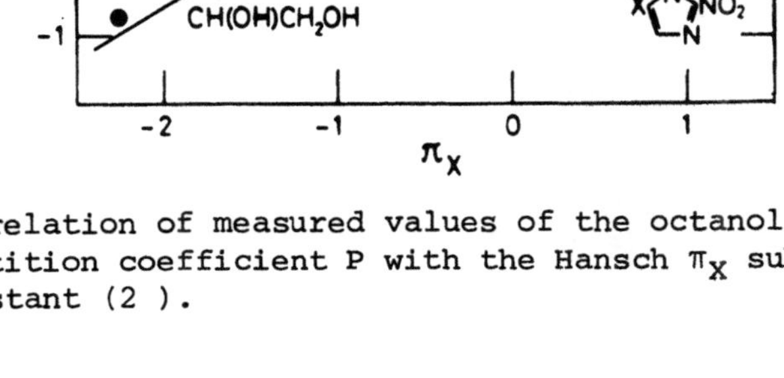

Fig. 4. Correlation of measured values of the octanol:water partition coefficient P with the Hansch π_X substituent constant (2).

It was concluded that the magnitude of P for substituted nitroimidazoles is influenced by (i) the position of the NO_2 group in the ring and (ii) by other substituents. For nitroimidazoles with identical sidechains, R^1 at N-1 ($R \neq H$) the lipophilicities P for 2-, 4- and 5-nitrosubstituents are in the approximate ratio 1:0.5:1.8 respectively.

It seems that the lipophilicity due to the $-NO_2$ group may be important for tissue penetration in regard to radiosensitizing effects but it does not seem to be important for the trichomonacide effect (25).

Acid/base equilibria are much influenced by the position of the NO_2 group in the molecule. This group, because of its electron affinity increases the acidic properties of the molecule.

Substitution in position 1 changes the basicity of the compounds in regard to loss of the acidic proton. Isomers of 4- and 5-nitroimidazole derivatives show a difference of basicity in the pKa range 2-2.5 (26). Until now no general correlation has been found between pKa and trichomonal or sensitizing activity. Studies are in progress on the effect of lipophilicity and pKa on the neurotoxic properties of nitroimidazoles (26).

REFERENCES

1. Source: J. R. Valentine, in: "Current Aspects of Exobiology", G. Mamikunian and M. H. Brigs, eds., Pergamon Press, London (1965), pp.1-12.
2. Source: "FLAGYL (Metronidazole) in Anaerobic Infections", Report by May & Baker Ltd., London, first edition (1977), p.10.
3. J. Stradins and S. Hiller, Izv. Akad. Nauk. Latv. SSR, 10:121 (1958). (Source: ref. 11 below).
4. J. Stradins, S. Hiller and Yu. Yur'ev, Doklady Akad. Nauk SSSR, 129:316 (1958).
5. J. Stradins and S. Hiller, Tetraedron, 20:Suppl. I. Nitro-compounds, 409 (1964).
6. J. Stradins, G. Reihamanis and R. Gavar, Elektrokhimiya 1:935 (1965); 3:178 (1976).
7. G. Reihmanis and J. Stradins, Izv. Akad. Nauk, Latv. SSR, Ser. Khim. p.23 (1967).
8. J. Stradins and G. Reihmanis, Izv. Akad. Nauk, Latv. SSR, Ser. Khim., p.377 (1969).
9. R. Gavar, J. Stradins and S. Hiller, Doklady Akad. Nauk, SSSR, 157:1424 (1964).
10. J. Stradins, R. Gavars, V. Grin and S. Hiller, Teoret. Eksper. Khimiya, 4:774 (1968).

11. J. P. Stradins, S. A. Hiller, R. A. Gavars, G. O. Reihmanis and L. H. Baumane, Electrochemical properties and biological activity of nitrofurans, in: "Biological Aspects of Electrochemistry", Experienta Suppl. 18:607 (1971).
12. G. E. Adams and D. L. Dewey, Hydrated electrons and radiobiological sensitization, Biochem. Biophys. Res. Comm. 12: 473 (1963).
13. J. F. Fowler, G. E. Adams and J. Denekamp, Radiosensitizers of hypoxic cells in solid tumours, Cancer Treatment Review 3: 227 (1976).
14. A. Breccia, G. Berrilli and S. Roffia, Chemical radiosensitiza- of hypoxic cells and redox potentials, Int. J. Radiat. Biol., 36:85 (1979).
15. J. D. Chapman, J. H. Raleigh, J. Pedersen, J. Ngan, F. Y. Shun, B. E. Mecker and R. C. Urtasun, Three roles for hypoxic sensitizers, Proceedings of VI International Congress of Radiation Research, Rad. Res. (1979).
16. A. Breccia,"Concepts of radiosensitization and techniques of investigation in radiosensitizers of hypoxic cells", A. Breccia, C. Rimondi and G.E. Adams, eds., Elsevier, p.1 (1979).
17. A. Breccia, S. Roffia, E. Gattavecchia, A. Borgatti, Studi elettrochimici e di radiolisi sulla radiosensibilizzazione delle cellule in stato ipossico, Chimica e Industria, 62: 258 (1980).
18. S. Roffia, E. Gattavecchia and A. Breccia, Electrochemical and γ-radiolysis studies on radiosensitization of hypoxic cell, Joint Meeting of Biophysical Society, New Orleans, June 1980.
19. S. Roffia, E. Gattavecchia and A. Breccia, Poster session on "Electrochemical investigation on the radiosensitization of hypoxic cells", Conference on Radiation Sensitizers and Protectors, Key Biscayne, USA, October 1979.
20. L. S. Matheson, L. M. Dorfam, "Pulse Radiolysis", The MIT Press, 1969.
21. G. E. Adams, I. R. Flockart, C. E. Smithen, I. J. Stratford, P. Wardman, and M. E. Watts, Electron-affinic sensitization VII: A correlation between structures, one electron reduction potentials, and efficencies of nitroimidazoles as hypoxic cell radiosensitizers, Radiat. Res. 67:9 (1967).
22. G. C. Lancini, J. Med. Chem., 12:775 (1969).
23. B. Blazevic, D. Karbah, V. Suvijic and F. Kayfez, Anti-tritocomonal activity of 2-, 4- and 5-nitroimidazoles and their physico-chemical properties, in: Proc. of Int. Cong. on Antimicrobial, p.389 (1971).
24. P. Wardman, The chemical basis for the development of hypoxic cell radiosensitizers, in: "Radiosensitizers of Hypoxic Cells", A. Breccia, G. E. Adams and C. Rimondi, eds., Elsevier (1979).

25. V. Suvijic, F. Kasfez, D. Kolbach and N. Blazecic, Physikalisch-chemische Eigenschaften und antitrichomonale wirkung von 2-, 4- und 5-Nitroimidazolen, Die Pharmazie, 3:131 (1972).
26. N. Blazevic, F. Kayfez and V. Suvijic, σ Values of some nitroimidazoles, J. Heterocy. Chem., 7:227 (1970).

We are obliged to Drs. S. Roffie, A. Grassi and E. Gattavecchia for their help in performing the experimental part cited in this paper.

Financial support was given by C.N.R. through the finalized project "Controllo della crescita neoplastica", research contract N. 80.01486.96.

IN VIVO TESTING OF RADIOSENSITIZERS

Juliana Denekamp

Gray Laboratory of the Cancer Research Campaign
Mount Vernon Hospital
Northwood, Middlesex HA6 2RN, England

Most anticancer drugs are designed to exploit the more rapid proliferation of tumour cells than of normal cells, and are cytotoxic in certain phases of the cell cycle. Nitro-imidazoles as tumour radiosensitizers are quite different since they have an equal sensitizing action on cycling and non-cycling cells, and on cells at any stage in their cell cycle. Their tumour-specific action relates to the presence of a large fraction of hypoxic cells in tumours, and not in normal tissues.

Hypoxic cells develop in tumours because of the unbalanced growth of the tumour cells and of the vascular components needed to provide an adequate blood supply. The production of new capillaries is inadequate, both in number and distribution; tumour cells become deprived of oxygen (and other nutrients) as the intercapillary distance becomes greater than 200-300 μm. This results from the rapid utilization of oxygen by the cells lying in a cuff around each capillary. The effect is illustrated in Figure 1.

In many tumours cords of viable cells are observed around capillaries, with necrotic areas lying at a distance of 90-150μ from the vessels (1). Thus tumour cells will die of their hypoxia in an untreated tumour and might be considered irrelevant. However, this hypoxia makes them extremely resistant to radiation (by a factor of about three). If a tumour is irradiated, the oxygenated radiosensitive cells adjacent to the capillaries are most readily killed. When they die, the oxygen diffusion path-length increases and the doomed hypoxic cells are rescued by a process of "reoxygenation", so that they can act as a nucleus of surviving cells from which tumour regrowth can occur.

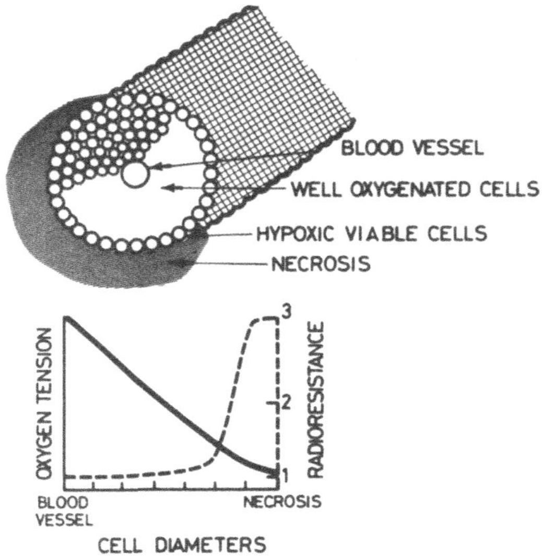

Fig. 1. Schematic representation of the cuff of viable cells around a tumour capillary. The oxygen tension falls in successive cell layers as it is utilized. When it reaches a critical concentration the cells become radio-resistant. This is believed to occur in 1-2 layers adjacent to areas of necrosis.

The electron-affinic radiosensitizers, including the nitro-imidazoles, can mimic oxygen by interacting with radicals at the site of radiation damage and physico-chemically fixing the damage. They do not replace oxygen in the metabolic pathways however, and can therefore diffuse through the cords of metabolising cells around each capillary, and can reach the distant hypoxic cells. The radiosensitizing action of the nitro-imidazoles occurs <u>only</u> in hypoxic cells; it is not observed when oxygen is present to exert its own radiosensitizing action. In general the effectiveness of the nitro-imidazoles as sensitizers is lower than that of oxygen, but their advantage lies in their ability to diffuse to poorly vascularised hypoxic regions.

The systems in which radiosensitization has been shown to occur are listed in Table 1. In all cases the effect has been demonstrated to be specific for cells deprived of oxygen, i.e. hypoxic cells.

The hypothesis that the radiosensitizing action of a number of nitro-containing, heterocyclic, organic compounds resulted from their electron affinity was first proposed by Adams and Dewey in 1963 (2). Since then hundreds of compounds have been shown to be

Table 1. Test Systems Which Have Shown Radiosensitization

In Culture	Hypoxic Bacteria
	Hypoxic Mammalian Cells
	Hypoxic Broad Bean Roots
In Mice	Artificially Hypoxic Skin
	Artificially Hypoxic Tumours
	Naturally Occurring Hypoxic Cells in Tumours (Single Doses)
	Naturally Occurring Hypoxic Cells in Tumours (Fractionated Doses)
In Man	Artificially Hypoxic Skin
	Subcutaneous and Pulmonary Tumour Nodules (Single Doses)
	Glioblastoma Multiforme (9 Fractions)

effective on bacterial cells; dozens of these are also effective on mammalian cells *in vitro* (3). For *in vivo* use however the constraints are much greater. The compounds must be non-toxic to the animal, must not bind to serum proteins or be rapidly metabolised, must be water soluble and freely diffusible through extracellular fluid and through cell membranes. These additional requirements have limited the number of compounds found to be useful *in vivo*. The group fulfilling these characteristics best seems to be the nitro-imidazoles, including *metronidazole (a 5-nitro-imidazole), and its 2-nitro analogue *misonidazole. These compounds were first tested in the early 1970's. Willson had recognised from the existing preclinical and clinical toxicology data, that lack of toxicity associated with Flagyl might enable this compound to be administered as a sensitizer *in vivo* if it was as electron affinic as its chemical structure suggested. Table 2 shows the speed with which the first *in vitro* and *in vivo* laboratory studies led to clinical testing of both metronidazole and misonidazole as adjuncts to radiotherapy of human cancer. Since then many other compounds have been tested in mice but as yet no more effective or less toxic substitute for misonidazole has been found.

The first *in vivo* experiments were performed on a normal tissue (mouse skin), made *artificially* hypoxic for the irradiation. This system was chosen because it forms a half-way step between *in vitro* studies and studies on tumours. It enables quantitative estimates

---

* metronidazole = Flagyl (May & Baker Ltd.)
misonidazole (Ro-07-0582) (Roche Ltd.)

Table 2. Time Scale of Hypoxic Cell Radiosensitizer Development

Year	*In vitro* Mammalian cells	Artificially hypoxic skin in mice	*In vivo* Solid tumours in mice	Clinical trials
1970	NDPP (2.1)	---	---	---
1971	Nitrofurans (2.0)	NDPP (1.1)	---	---
1972	Metronidazole (1.9)	Metronidazole (1.6)	NDPP (1.1)	Metronidazole (low dose) Edmonton, Canada
1973	Misonidazole (2.5)	Misonidazole (2.2)	Metronidazole (1.6) NDPP (1.3) Misonidazole (1.6-2.1)	Metronidazole (high dose) Northwood, U.K.; Edmonton, Canada
1974	---	---	Misonidazole (1.5-2.3) 8 tumour systems, single doses	High single doses Misonidazole (1.2) Northwood
1975	---	---	fractionated schedules (1.1-1.6)	Northwood, Edmonton Vienna, Göteborg, Cape Town, Toronto
			Misonidazole + neutrons	Controlled clinical trials in USA and UK

The sensitizer enhancement ratios obtained are shown in brackets.
SER = the radiation dose without drug divided by that with drug to achieve the same level of effect (see ref. 4).

of the survival of individual epithelial cells to be made with great precision, and is fairly rapid, yielding results within three weeks (5,6). Defined areas of epithelium are isolated from the surrounding epidermis by a "moat" of heavily irradiated cells and then graded doses of radiation are given to these test areas and the number of islands regrowing as microscopically visible clones are scored from 12-21 days after irradiation (Fig. 2).

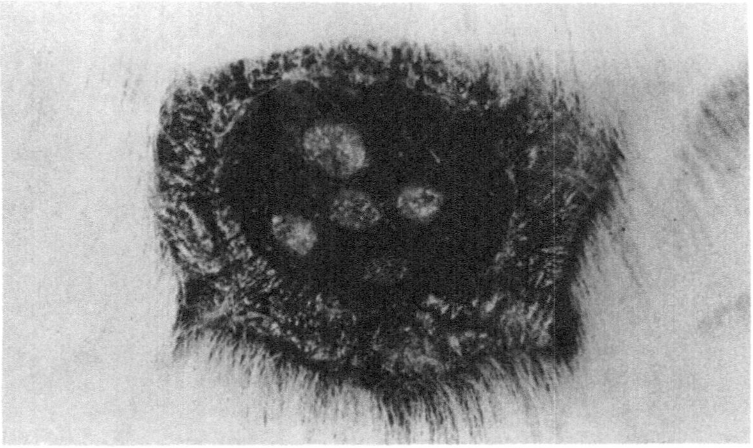

Fig. 2. Five epidermal clones regrowing into macroscopically visible islands at about 2½ weeks after irradiation.

The depletion of epidermal clones is observed in mice breathing oxygen at doses below 2 krad whereas if the mice are made hypoxic (by breathing nitrogen for 35 seconds) the skin becomes 2.7-2.9 times more resistant and 5-6 krad are needed to eliminate most of the epidermal cells. Figure 3 shows the response of mice irradiated in oxygen or nitrogen, either without a radiosensitizer or after administration of metronidazole or misonidazole.

No radiosensitization was observed in mice breathing oxygen, or if the drug was administered 5 minutes after irradiation. If the drug was administered before irradiation, however, the degree of sensitization of hypoxic cells was considerable. With both drugs it increased with increasing drug dose and was more effective with misonidazole than with metronidazole.

From a series of such experiments the radiosensitizing action of these two drugs (and several others) was determined as a function of drug dose administered. Fig. 4 shows the comparison. Misonidazole (0582) was more effective than metronidazole (Flagyl), and 75% of the oxygen effect could be achieved at doses below the LD_{50} value. Two other compounds, NDPP and nifurpipone, were similarly effective but their use was limited by their toxicity.

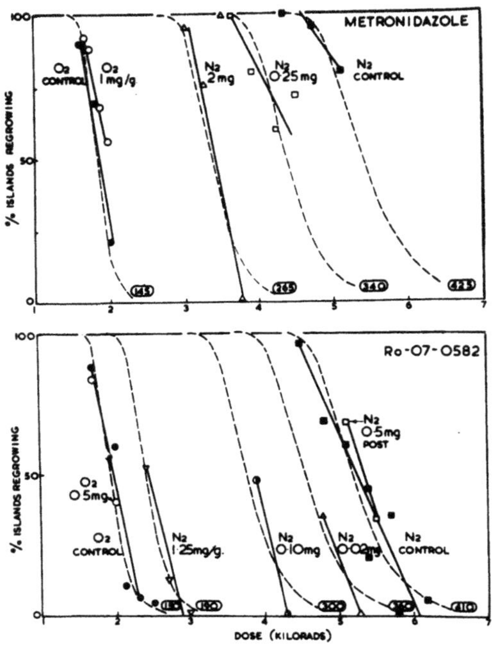

Fig. 3. The percent of epidermal islands regrowing as a function of dose. The inhaled gas and the drug dose administered (in mg/g body weight) are indicated against each line. Misonidazole (Ro 07-0582) is clearly more efficient than metronidazole. (Redrawn from ref. 7.)

These animal experiments were soon followed by Phase I and II clinical studies. In the study of Dische et al.(8) a similar model using hypoxic human skin was devised in order to determine whether sufficiently high drug concentrations could be achieved clinically to produce sensitization of artificially hypoxic cells. Small areas of skin on a patient's limb were irradiated with a strontium β-emitting plaque, either under conditions where the skin was well oxygenated, or where it was made hypoxic by the application of a tourniquet and by surrounding the limb in an atmosphere of flowing nitrogen. Nine patients were treated in this way and the degree of reddening of the areas of skin was scored over the following 3 months. Dose response curves were obtained and enhancement ratios were derived for several different concentrations of metronidazole and misonidazole. These clinical results are compared with the mouse skin data in Figure 5. It is clear that the effects in mouse and in man are quantitatively similar. More than 50% of the full sensitization achievable with oxygen was achieved with doses of misonidazole that were tolerated in patients as a single administration.

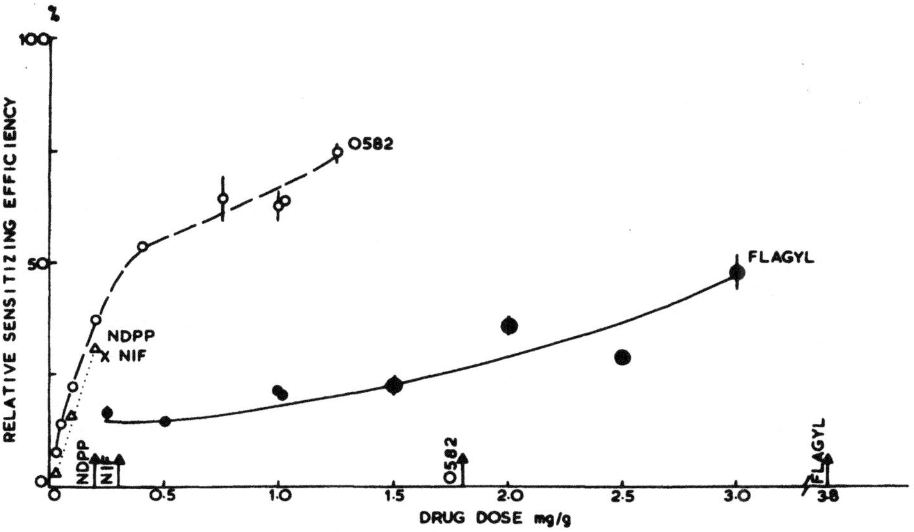

Fig. 4. The sensitizing efficiency of four compounds expressed as a percent of the efficiency of oxygen. The single dose acute LD_{50} values are indicated by arrows. (From ref. 7)

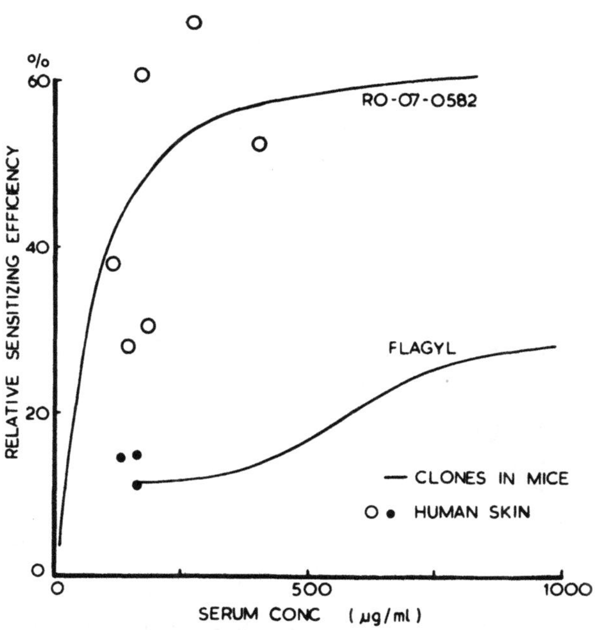

Fig. 5. The sensitizing efficiency of misonidazole and metronidazole expressed as a percentage of the efficiency of oxygen. The lines represent the data from hypoxic mouse skin. The points represent the results from human skin made artificially hypoxic for the irradiation. (Data from refs 7,8).

Although it was important to demonstrate the feasibility of radiosensitization of hypoxic cells in artificially hypoxic tissues, the important question was, of course, whether such radiosensitization could be demonstrated in tumours, particularly in the naturally-occurring hypoxic cells believed to exist at 6-10 cell diameters from the nearest capillary. To this end many studies were undertaken at the Gray Laboratory and elsewhere, on a wide variety of animal tumours. Dose response curves for increasing tumour damage with increasing dose were constructed by scoring a variety of endpoints. These included regrowth delay, the percentage of locally controlled tumours, the loss of incorporated radioactivity or the survival of tumour cells when excised and assayed quantitatively <u>in vitro</u> or in recipient mice. All of these studies have demonstrated that misonidazole and metronidazole are effective radiosensitizers, with the 2-nitroimidazole being more effective than the 5-nitro compound.

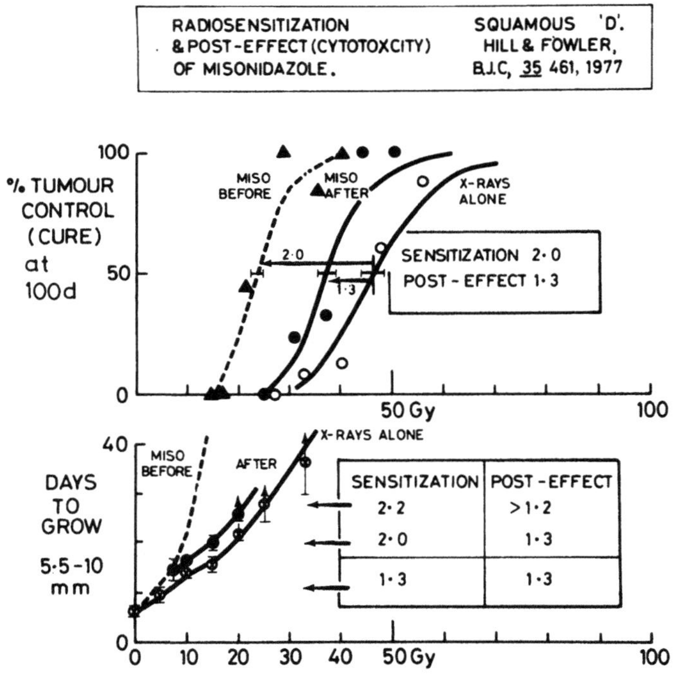

Fig. 6. Regrowth delay (lower panel) and local tumour control (upper panel) of a mouse squamous carcinoma. Curves are shown for animals treated with X-rays alone, with misonidazole given before irradiation (SER 2.0-2.2), or given after irradiation to demonstrate the hypoxic cell cytotoxic action of the drug. (Date from ref. 9).

An example of regrowth delay and cure data from the same set of tumour experiments is shown in Figure 6. At low radiation doses a small delay in the time taken for the tumour to reach a certain size is observed; this delay increases with increasing radiation dose and eventually some of the tumours fail to regrow at all, i.e. are local cures. Thus at the higher doses the percent of locally controlled tumours can be scored as a function of dose.

The cure data and the upper part of the regrowth delay data demonstrate that a large degree of radiosensitization of the naturally occurring hypoxic cells is achieved with a large drug dose (1 mg/g body weight of mice). At the bottom end of the regrowth delay curves the sensitization is less marked. In this region the tumour response is mainly due to the radiation induced killing of the well-oxygenated cells forming the cuff around each capillary, and these cells are not sensitized. Thus it is only when the hypoxic radio-resistant cells begin to dominate the response that marked sensitization is observed. This is exactly what would be predicted from a consideration of the survival curves for a mixed population

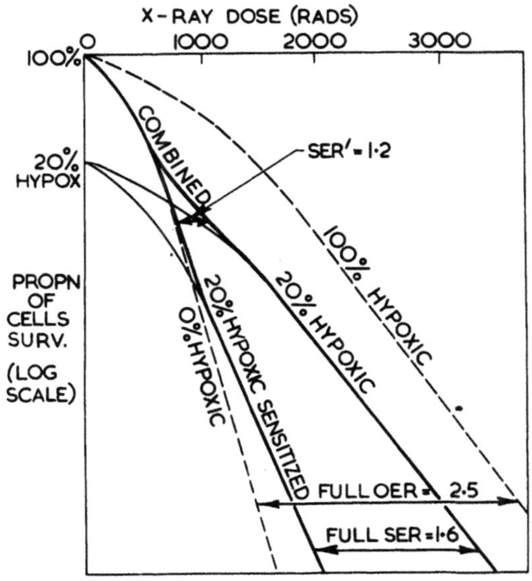

Fig. 7. Schematic representation of the survival curves for oxygenated or hypoxic cells, or for a mixed population containing 20% hypoxic cells. If a sensitizer giving a full SER of 1.6 is used together with 1000 rads of radiation, a smaller SER' of only 1.2 will be observed because of the masking effect of the oxic cells.

of well-oxygenated and hypoxic cells, as illustrated in Figure 7. Because the radiosensitizing action of the drug is limited to the hypoxic sub-population it is only observed when the response of this sub-population is expressed.

Figure 8 shows a comparison of the regrowth delay (with and without misonidazole) in tumours that have been clamped to occlude their blood supply and render all the cells hypoxic, or have been irradiated under normal aerobic conditions. In the clamped tumours a large degree of sensitization is observed at all dose levels, being similar at high and low doses. This, compared with the curve for unclamped tumours, again illustrates the singular effectiveness of these radiosensitizers against hypoxic, and not oxic, cells.

One of the most complete sets of data on the radiosensitizing action of nitro-imidazoles on a mouse tumour comes from Sheldon and Hill (10). Two sets of their early data are shown in Figure 9, where the percentage of MT tumours locally controlled by X-rays alone or after administration of various doses of metronidazole or misonidazole are illustrated. The increasing sensitization with increasing dose is clearly demonstrated, and the greater effectiveness of the 2-nitro-imidazole is seen here, as in the skin and in the mammalian systems tested *in vitro*.

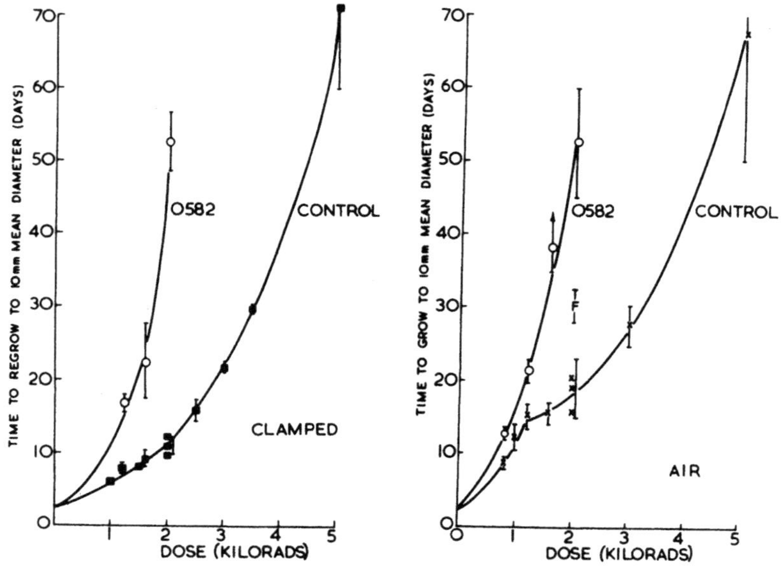

Fig. 8. Regrowth delay as a function of dose for the mouse mammary carcinoma MT. The left hand panel shows tumours made artificially hypoxic by clamping. The right hand panel shows the biphasic curve characteristic of a mixed oxic and hypoxic population. (Data from ref. 11).

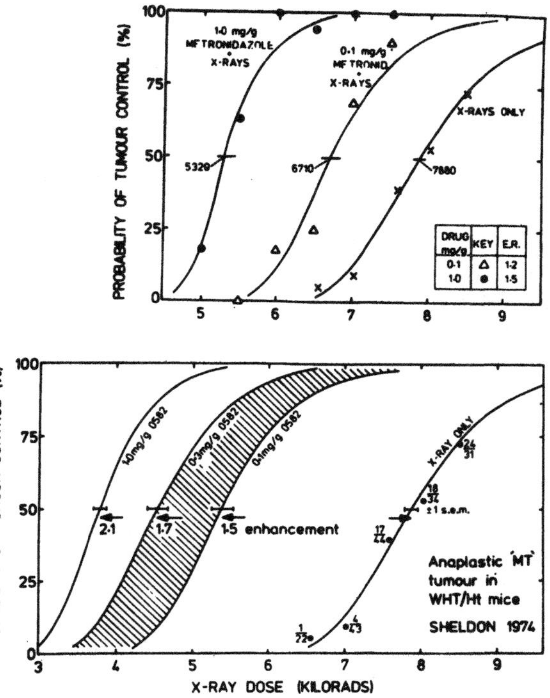

Fig. 9. Local control of the anaplastic MT tumour after irradiation with X-rays alone or after the stated doses of metronidazole or misonidazole (0582). The hatched area in the lower diagram indicates the sensitization that should be achievable with large single doses in patients. (Data from ref. 10.)

Since the effectiveness of the hypoxic cell sensitizer depends upon the presence of hypoxic cells in tumours it seems pertinent at this point to ask whether hypoxic cells exist in all tumours, and whether they have been demonstrated in man, or are simply inferred from the corded structure illustrated in Figure 1, which is sometimes seen in both human and murine tumours. Table 3 is a summary of the hypoxic fraction estimates for a wide range of animal tumours and for a few human tumours. Hypoxic cells clearly exist in significant proportions in most tumours; these cells are radioresistant and may limit the local control that can be achieved clinically.

This uniformity of tumours in terms of the presence of such hypoxic cells is illustrated in Figure 10, where the results of many studies of sensitization as a function of drug concentration have been summarised. For single doses of radiation and of drug the degree of sensitization increases with increasing drug dose, to between 1.7 and 2.4 in different tumour systems at a dose of 1 mg/g.

Table 3. Hypoxic fraction estimates in mouse and human tumours.

		Sarcomas %	Carcinomas %
Mouse	a) Published estimates	$\geq$ 50	21
	reviewed by Fowler and	14	0.2
	Denekamp (ref. 4)	15	18
		14	20
		17	5
		12	19
		35	1
		$\leq$ 10	17
		50	6
		1	$\leq$ 1
			$\geq$ 46
			5
			1
	Gray Laboratory tumour	0-0.01	10-30
	sublines (12)	1-10	2-25
		4- 8	7-15
		10-30	5-20
	The range of values comes	1- 3	
	from different methods of	5-20	
	estimating hypoxic	30-70	
	fractions	0-20	
		23-50	
		5-25	
Human	Data of		
	Thomlinson et al., (13)	$\geq$ 12	12-20
	and		$<$ 1
	Ash et al., (14)	1-20	50-70
	analysed as described	50-80	10-35
	in Denekamp et al., (15)	1	

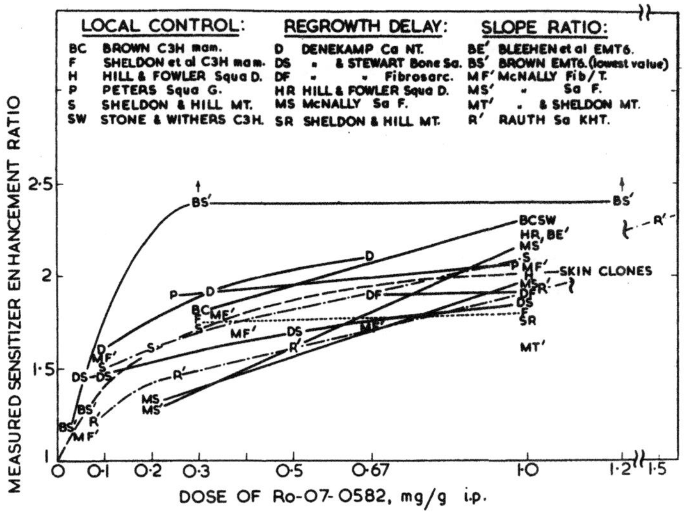

Fig. 10. Enhancement ratios for misonidazole measured by three different types of assay in a wide variety of different tumour types. (For references see ref. 4).

Although sensitization has been demonstrated in most animal tumours (see ref. 4 for review) the effect on human tumours is, of course, much more difficult to quantitate. A number of studies of regrowth delay in patients with multiple subcutaneous tumour nodules have been performed by Thomlinson et al. (13) and by Ash et al. (14). In these studies sensitization by a factor of 1.2-1.6 has been observed with single doses of radiation and large drug doses. Thus, although mouse tumours differ from human tumours in many respects (Table 4), the similarity of the critical diffusion path-length for oxygen through metabolising tissue makes mouse tumours a good model for predicting the response to electron-affinic radiosensitizers.

The presence of hypoxic cells and the ability of nitro-imidazoles to diffuse to them in adequate concentrations to achieve radiosensitization has been clearly demonstrated. However the drug doses needed are high, and cannot be used repeatedly in man because of the troublesome peripheral neuropathy encountered if a dose in excess of 12 g/m^2 is administered over the course of 4-6 weeks (16). Fractionating the drug dose does not diminish its neurotoxicity, although it does diminish its radiosensitizing action if reoxygenation occurs between the repeated small doses. This leads the clinician to ask how he should best use misonidazole, and whether a better drug is just around the corner. The approaches that are currently being used in the controlled clinical trials around the world range from the use of unconventional radiation schedules, e.g. 6 large doses of 600 rads in 3 weeks, each accompanied by a

Table 4. Proportionality of Tumour Models.

		Mouse	Man	Ratio
Weight	Body wt.	20-40 g	50-100 kg	X 2,500
	Tumour wt.	0.1-1.0 g	0.01-1.0 kg	X 10-1000
	Tumour wt. / Body wt.	1%	0.1%	X 0.1
Time	Host's life span	2.5 years	75 years	X 30
	Tumour T_D	1-10 days	20-100 days	X 10-20
	Tumour T_C	12-14 hours	24-80 hours	X 2
Size	Tumour mean diameter	1 cm	10 cm	X 10
	Microscopic distance blood vessel to necrosis	60-100 μm	100-150 μm	X 2

large drug dose of 2 g/m^2, to using small doses of 0.4 g/m^2 with each of the conventional 30 fractions of 200 rads given over a six-week period. The question of which schedule is likely to produce the best clinical result has recently been tackled in a computer simulation of several proposed treatment schedules (17). The most advantageous schedule will depend to some extent on the survival curve characteristics of the human tumour cells, and to a large extent on their reoxygenation pattern. Unfortunately, information is lacking on both these aspects. In general, a larger degree of sensitization will be observed with a few large fractions, each with a large drug dose. However, there may be disadvantages of changing to an unorthodox fractionation scheme which may counteract this greater sensitization. In many of the simulations the most effective treatment was that using conventional schedules (each fraction with a small drug dose) even though the sensitization observed was then much smaller.

The time of administration of the drug before irradiation is also of importance, since an adequate time for absorption, distribution and diffusion to the hypoxic cells must be allowed, but not so long that the drug concentration is diminished because of metabolism or excretion. In the mouse, the half life of misonidazole is about 1 hour (18) and the optimal interval is 36-60 minutes (Table 5). In man the half life is much longer (10-18 hours), and the peak tumour concentration occurs at 2-4 hours. It has therefore become an accepted convention to irradiate 3-4 hours after oral administration of the misonidazole. This timing is based on HPLC estimates of drug concentrations in human tumours, although these represent concentrations in the bulk of the tumour and may not be an accurate reflection of the concentration in the critical hypoxic cells.

The search for new drugs continues in the laboratory. Many other nitro-imidazoles have been identified which will give a similar degree of radiosensitization on an equimolar basis to that obtained with misonidazole. The choice of a better compound to replace misonidazole clinically will depend upon achieving a greater therapeutic index (i.e. less toxicity for the same sensitization) or a compound that is much more efficient (i.e. in which less drug will be needed). The toxicity aspect of this search has been hampered by the lack of a suitable animal model for the sensory peripheral neuropathy which is dose-limiting in the clinic. In mice, nerve conduction velocity experiments have demonstrated different toxicities of several compounds but have not produced the lesser toxicity with metronidazole than with misonidazole that has been observed in clinical practice (19,20). This may be because of the artefacts resulting from the drop in body temperature that accompanies large drug doses, or it may be because in this respect the mouse is a poor model for man. Other tests have included the swimming response of goldfish, the ability of chickens to walk up ladders, the ability of mice to stay on a rotating rod as it accelerates, or to walk along a narrowing plank, and the histochemical assessment of lysosomal damage in excised nerves. These latter two tests seem to be the most promising and will hopefully lead to a more rational ranking of potential new radiosensitizers than has been possible up to now using single dose LD_{50} values as the index of toxicity (21). The search for better sensitizers is currently focussed on designing nitro-imidazole molecules with a lower lipophilicity so that they will not readily enter nervous tissue, but will retain their electron affinic radiosensitizing properties (22).

The experience with misonidazole in the preclinical radiobiology has shown us that chemical synthesis, pharmacology, toxicology and radiobiology must proceed hand in hand. So far, the screening of potential compounds on mammalian cells *in vitro* has been a good guide to the radiosensitizing action that can be achieved *in vivo* (e.g. ref. 23), but a detailed knowledge of the toxicity, of the half life in serum and in tumour and of the optimum time for drug administration are needed before a better sensitizer

Table 5. The Importance of the Time Between Drug Administration and X-irradiation.

Tumour	Assay System	Drug Dose mg/g	Enhancement Ratio	Optimum time interval (min)	Reference
<u>Misonidazole</u>					
C3H Mam. Ca	Local control	1.0	2.1-2.3	10-30	Stone & Withers, 75b
EMT 6	Excision/*in vitro* survival	0.3	2.4-3.1	30-60	Brown, 75.
EMT 6	" " "	1.2	2.4-2.9	30-60	Brown, 75.
Rat Sa 180	^{125}IUdR loss	0.5	1.5	15	Porschen et al., 77.
Ca NT'	Regrowth delay	0.3	1.9	45	McNally et al., 78.
Anaplast MT1	Local control	0.2	1.5	45-60	" " "
Fib/T	Excision/*in vitro* survival	0.3	1.8	20-90	" " "
Fib/T	" " "	1.0	2.1	50-120	" " "
Bone Sa 2	Regrowth delay	0.5	1.7	30-60	" " "
Bone Sa 2	" "	1.0	1.8	15-30	" " "
<u>Metronidazole</u>					
EMT 6	Excision/*in vitro* survival	1.0	1.8-2.1	30-60	Brown, 75.
C3H Mam. Ca	Local control	0.1	1.2	3-30	Stone & Withers, 74.
C3H Mam. Ca	" "	1.0	1.5	30-60	Stone & Withers, 75a.

For details of references, see ref. (4).

can be identified with certainty.

1. R. H. Thomlinson and L. H. Gray, The histological structure of some human lung cancers and the possible implications for radiotherapy, Brit. J. Cancer, 9:539 (1955).
2. G. E. Adams and D. L. Dewey, Hydrated electrons and radio-biological sensitization, Biochem. Biophys. Res. Comm., 12:473 (1963).
3. G. E. Adams, E. D. Clarke, I. R. Flockhart, R . S. Jacobs, D. S. Sehmi, I. J. Stratford, P. Wardman, M. E. Watts, J. Parrick, R. G. Wallace and C. E. Smithen, Structure-activity relationships in the development of hypoxic cell radiosensitizers. I. Sensitization efficiency, Int. J. Radiat. Biol., 35:133 (1979).
4. J. F. Fowler and J. Denekamp, A review of hypoxic cell radio-sensitization in experimental tumors, Pharmacol. & Ther., 7:413 (1979).
5. H. F. Withers, The dose survival relationship for irradiation of epithelial cells of mouse skin, Brit. J. Radiol., 40: 187 (1967).
6. J. Denekamp and B. D. Michael, Preferential sensitization of hypoxic cells to radiation in vivo, Nature New Biol., 239:21 (1972).
7. J. Denekamp, B. D. Michael and S. R. Harris, Hypoxic cell radiosensitizers: comparative tests of some electron affinic compounds using epidermal cell survival in vivo, Radiat. Res., 60:119 (1976)
8. S. Dische, A. J. Gray and G. D. Zanelli, Clinical testing of the radiosensitizer Ro-07-0582. II. Radiosensitization of normal and hypoxic skin, Clinical Radiol., 27:159 (1976).
9. S. A. Hill and J. F. Fowler, Radiosensitizing and cytocidal effects on hypoxic cells of Ro-07-0582, and repair of X-ray injury, in an experimental mouse tumour, Brit. J. Cancer, 35:461 (1977).
10. P. W. Sheldon and S. A. Hill, Hypoxic cell radiosensitizers and local control by X-ray of a transplanted tumour in mice, Brit. J. Cancer, 35:795 (1977).
11. J. Denekamp and S. R. Harris, Tests of two electron-affinic radiosensitizers in vivo using regrowth of an experimental carcinoma, Radiat. Res., 61:191 (1975).
12. J. Denekamp, D. G. Hirst, F. A. Stewart and N. H. A. Terry, Is tumour radiosensitization by misonidazole a general phenomenon? Brit. J. Cancer, 41:1 (1980).
13. R. H. Thomlinson, S. Dische, A. J. Gray and L. M. Errington, Clinical testing of the radiosensitizer Ro-07-0582. III. Response of tumours, Clinical Radiol., 27:167 (1976).

14. D. V. Ash, M. J. Peckham and G. G. Steel, A quantitative study of human tumour response to radiation and misonidazole. Brit. J. Cancer, 39:503 (1979).

15. J. Denekamp, J. F. Fowler and S. Dische, The proportion of hypoxic cells in a human tumor, Int. J. Radiat. Oncol. Biol. Phys., 2:1227 (1977).

16. S. Dische, M. I. Saunders, P. Anderson, R. C. Urtason, K. H. Kärcher, H. D. Kogelnik, N. M. Bleehen, T. L. Phillips and T. H. Wasserman, The neurotoxicity of misonidazole. The pooling of data from five centres, Brit. J. Radiol. 51:1023 (1978).

17. J. Denekamp, N. J. McNally, J. F. Fowler and M. C. Joiner, Misonidazole in fractionated radiotherapy: little and often? Brit. J. Radiol., (in press) (1980b).

18. P. Workman, Effects of pretreatment with phenobarbitone and pehotoin on the pharmacokinetics and toxicity of misonidazole, Brit. J. Cancer, 40:335 (1979).

19. D. G. Hirst, B. Vojnovic, I. J. Stratford and E. L. Travis, The effect of the radiosensitizer misonidazole on motor nerve conduction velocity in the mouse, Brit. J. Cancer, 37: Suppl. III: 237 (1978).

20. D. G. Hirst, B. Vojnovic and B. Hobson, Changes in nerve conduction velocity in the mouse after acute and chronic administration of nitroimidazoles, Brit. J. Cancer, 39: 159 (1979).

21. C. Clarke, K. B. Dawson, P. W. Sheldon, D. J. Chaplin, I. J. Stratford and G. E. Adams, A quantitative cytochemical method for assessing the neurotoxicity of misonidazole, in: "Radiation Sensitizers: Their use in the Clinical Management of Cancer", L. W. Brady, ed., Masson Publishing Inc., New York. (1980).

22. J. M. Brown and P. Workman, Partition coefficient as a guide to the development of radiosensitizers which are less toxic than misonidazole, Radiat. Res., 82:171 (1980).

23. J. Denekamp, Testing of hypoxic cell radiosensitizers *in vivo*, Cancer Clinical Trials, 3:139 (1980).

NITROIMIDAZOLES AS HYPOXIA-MEDIATED DRUGS IN CANCER CHEMOTHERAPY

I. J. Stratford and G. E. Adams

Institute of Cancer Research
Clifton Avenue,
Sutton, Surrey, UK

INTRODUCTION

There is abundant evidence that human tumours, even of quite small size (2 mm) contain a significant proportion of cells that are deficient in oxygen (1). These hypoxic cells are radiation-resistant and are believed to limit in some instances the effectiveness of the treatment of cancer by radiotherapy. Several of the papers presented in this Conference have discussed how nitroimidazoles might be used to overcome this problem by specifically increasing the radiation sensitivity of the resistant hypoxic cells. This paper is concerned, however, with the potential use of nitroimidazoles in cancer chemotherapy.

There are reasons for suspecting that hypoxic cells may be resistant to some drugs often used in cancer chemotherapy. These cells tend to be located in relatively poorly-vascularised regions of tumours, hence they may be less accessible to cytotoxic drugs. Further, hypoxic cells are out of cycle or at least have their progression slowed down and hypoxia may substantially affect the activity of a given drug, particularly if oxygen-dependent energy requirements are to be met.

Cyclophosphamide has been reported to preferentially kill the oxic cells in a transplantable rat tumour (2) and a similar effect was noted for the action of BCNU against the B16 melanoma (3). Table 1 summarises the results of investigations of this effect in oxic and hypoxic mammalian cells treated with a variety of anti-cancer drugs.

Table 1. The cytotoxic effect of some commonly used cancer chemotherapeutic agents on aerobic and hypoxic mammalian cells *in vitro*

Drug	Response of acutely hypoxic compared to exponentially growing aerobic cells	References
Adriamycin	Sensitive	5
	No difference	4,7
	Resistant	6,9
Bleomycin	Resistant	5,6,8,10
Actinomycin D	No difference	4
	Resistant	5,11
Vincristine	No difference	4
	Resistant	5
5-Fluorouracil	No difference	4,5
Methatrexate	No difference	4,5
Procarbazine	Resistant	5
Streptonigrin	Resistant	5
BCNU	No difference	4,5,6
Cis-Pt(II)	Sensitive	12
	No difference	4,5,6
Mitomycin C	Sensitive	5
Thio-tepa	Sensitive	8
Melphalan	Sensitive	8

Cell lines used: Refs. 4,5: EMT6
6: CHO
7,8,10-12: V79
9: WHFIB

It is clear from the data in Table 1 that the gaseous environment in which the cells are treated can affect their response to some cytotoxic drugs. However, it is also apparent that, even for the same cell-line, results from different laboratories show different effects of hypoxia. This can be due to differences in experimental design since it is known that the degree and duration of hypoxia to which cells are exposed can profoundly affect their response to treatment with cytotoxic drugs. This is illustrated by the study by Smith et al. (7) of the effect of the duration of hypoxic exposure on cells treated with adriamycin. The cytotoxic effect of this drug on aerobic Chinese hamster V79 cells was the same as that when cells were rendered hypoxic for one hour before treatment with adriamycin under hypoxia (acutely hypoxic). However, when cells were rendered hypoxic for six hours or more, prior to treatment with the drug (chronically hypoxic) they were much more resistant to the cytotoxic effects. These data and some of the responses summarised in Table 1, indicate therefore that in tumours, the presence of hypoxic cells can influence overall tumour response to cytotoxic drugs. If this is so, it follows therefore that an improved anti-tumour effect could be achieved if chemotherapeutic agents were used in combination with a compound, or compounds, which are specifically cytotoxic towards hypoxic cells.

CYTOTOXICITY OF NITROIMIDAZOLES

It is now well-established that many nitroimidazoles are considerably more toxic to mammalian cells in hypoxia than they are under oxic conditions. This has been demonstrated in various single-cell cultures _in vitro_ (13-16), in cultures of multicellular spheroids (17-19) and _in vivo_ (20,21). It has been demonstrated both in experimental animals and in man that misonidazole in some respects has favourable pharmacological properties for the study of interactions of drugs with hypoxic cells. It is relatively stable and does not undergo protein-binding. For these and other reasons, misonidazole is able to diffuse efficiently through poorly-vascularised regions of tumour tissue and can therefore reach hypoxic cells. It has been shown, for example, that the distribution of misonidazole in large human tumours is such that drug concentrations in regions where hypoxic cells are expected to occur, are often as high as the concentrations in blood plasma (22).

CHEMOPOTENTIATION BY NITROIMIDAZOLES

General background

In view of its properties, misonidazole has been selected by most authors for studies of the combined cytotoxic effects of nitroimidazoles and other anti-cancer drugs. Initially, the rationale behind these studies was based simply on the assumption that the differential hypoxic cytotoxicity of misonidazole would lead to a greatly-enhanced anti-tumour effect by removing the sub-population of cells (hypoxic) that might be resistant to attack by the second drug. It has now been amply demonstrated that such enhanced anti-tumour effects occur in a variety of experimental tumours treated with combinations of misonidazole and one of a series of anti-cancer drugs, particularly alkylating agents. It is now clear that these phenomena cannot be explained entirely in terms of the _separate_ cytotoxic actions of each drug with different cellular sub-population in the tumour, i.e. oxic and hypoxic cells. Rather, there is substantial evidence that misonidazole and other nitroimidazoles actually potentiate the cytotoxic action of the second drug.

Results of experiments in vivo were first reported by Rose et al. (23) and by Clement et al. (24). These authors reported that misonidazole substantially enhanced the anti-tumour effects of cyclophosphamide (CY), melphalan, methyl CCNU, aziridinyl-benzoquinone and 5-fluorucacil. These early studies also showed that the increased anti-tumour effect was not accompanied by a corresponding increase in normal tissue damage indicating some degree of tumour selectivity.

Interaction of misonidazole and cyclophosphamide in vivo

Table 2 includes data showing the potentiating effects of misonidazole on some experimental tumours in mice treated with CY. In these studies which employed either clonogenic cell-assay or regrowth delay as end-points, enhancement ratios up to 2.6 have been observed. However, clearly the magnitude of this potentiating effect depends upon several factors. These include a) tumour type, b) tumour size, c) timing between administration of misonidazole and the other cytotoxic agent, and d) the concentration of misonidazole.

The studies of Clement et al. (24) utilised four different tumours. The M5026 and the Lewis lung tumours were implanted subcutaneously (sc) and treated when they had grown to a size of between 8 and 9 mm in diameter. The B16 tumours were implanted intramuscularly (im) and treated 6 days later before they had become palpable. Further, the drug combination was tested in the L1210 leukaemia system growing as ascites tumours. The greatest enhancement was seen for the M5026 tumour but much less was seen in the Lewis lung system. No potentiation was observed either in the early B16 or in the L1210 tumours.

Other studies with the Lewis lung tumour (23,27) have shown enhancement ratios of ≥2.0 and in the same series of investigations even greater potentiation was observed on a human pancreatic tumour growing in immune-deprived mice (27). Variation in the magnitude of potentiation of CY by misonidazole has also been reported by Twentyman (30) who showed an enhancement ratio of 2.0 in the RIF1 tumour but substantially less in the KHT tumour.

Variations in chemopotentiation have also been observed in two human melanoma xenografts treated with melphalan and misonidazole. Tumours HX34 and HX47 showed enhancement ratios of 1.3 and 2.0 respectively when compared with the responses to melphalan alone (33).

Table 2. Enhancement by misonidazole of the anti-tumour effect of cyclophosphamide (CY)

Dose range mg/kg	Tumour	Assay	MISO dose mg/g	Timing MISO → CY	Maximum enhancement	Reference
Up to 80	Lewis lung	_in vitro_	1.0	simultaneous	2.2 2.0	23 27
Up to 120	HX32	_in vitro_	1.0	simultaneous	2.6	27
Up to 175	WHFIB	_in vitro_	1.0	1 hour	2.0	28
Up to 75	RIF 1	_in vitro_	0.75	30 minutes	1.5	29
Up to 300	M5026	regrowth delay	1.0	1 hour	1.8	24
Up to 300	Lewis lung	regrowth delay	1.0	1 hour	<1.2	24
Up to 75*	RIF 1	regrowth delay	0.75	30 minutes	2.1	29
Up to 50*	RIF 1	regrowth delay	1.0	simultaneous	2.0	30
Up to 140	Lewis lung	regrowth delay	1.0	simultaneous	2.0	27

* Dose-modifying up to these dose levels. At higher doses the enhancement ratio decreases.

Effect of tumour size

For a given tumour type, the size at the time of treatment influences the magnitude of the response to combination treatment. For example, the Lewis lung tumour shows a greater response to melphalan and misonidazole when the tumour is grown as a large intramuscular tumour (0.2-0.4 g) compared with that seen when the tumours are treated as small lung colonies. In similar experiments Martin et al. (28) used the combination of CY and misonidazole to treat the WHF1B tumour either implanted sc on the chest of mice (6-8 mm diameter) or as 2 mm lung tumours. The enhancement of CY damage was 2.0 in the large tumours but only 1.4 in the small lung nodules. However, this is not a general phenomenon since Tannock (26) did not observe any difference in response to misonidazole plus CY in both large (0.2-0.3 g) and small (non-palpable) KHT tumours.

Effect of drug-dose and timing

It has been clearly demonstrated that radiosensitization by nitroimidazoles *in vivo* is strongly dependent on the interval between drug administration and irradiation (35,36). The dependence of the extent of "chemosensitization" by nitroimidazoles on timing has also been demonstrated (23,28,29,31). Generally, the optimum sequence is for misonidazole to be given up to 1 hour before administration of the cytotoxic agent, although significant enhancements have still been observed when misonidazole is given up to 6 hours before CY (28) or up to 6 hours after BCNU (31).

The magnitude of the enhancement by misonidazole of damage caused by cytotoxic agents can also depend upon the dose of misonidazole. Typical data showing such dependence are given in Fig. 1 (34), where the response of the anaplastic MT tumour to melphalan has been assessed using the soft agar cloning technique. Tumours were treated at a mean diameter of 6-7 mm and mice were injected ip with either misonidazole or Ro 05-9963 (desmethylmisonidazole) dissolved in DMSO immediately before administration of 5 μg/g melphalan. No significant toxicity was seen for misonidazole or Ro 05-9963 alone but cell survival was reduced to 3.6×10^{-2} for melphalan alone. However, when the nitroimidazoles were administered with the melphalan, tumour cell survival was greatly reduced. For 1 mg/g misonidazole the enhancement ratio is about 2 and even at one-tenth of this dose (100 μg/g), an enhancement ratio or dose-modifying factor (DMF) of 1.5 is still evident. This suggests that potentiation may be achievable with misonidazole at doses within the clinical range.

A similar concentration dependence was observed by Law et al. (29) for the combination of CY and misonidazole and by Siemann (32)

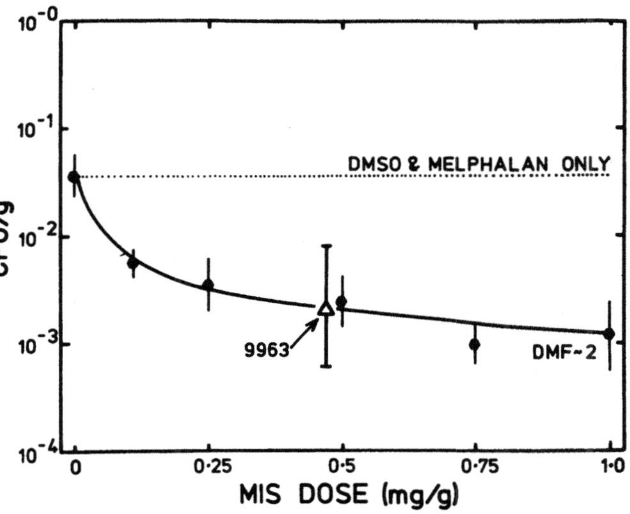

Fig. 1. Potentiation by misonidazole (or Ro 05-9963) of the cytotoxic effect of melphalan in MT tumours in WH mice (34).

for the combination of CCNU and misonidazole. Both these studies were carried out with the RIF 1 tumour. In contrast, other work with CY (28,30) showed that the enhancing action of misonidazole is reduced considerably as the dose of nitro compound is decreased.

Nitroimidazoles other than misonidazole have been shown to potentiate the cytotoxic action of some alkylating agents in vivo (25,26,28,29,31). Data for desmethylmisonidazole are shown in Fig. 1 indicating the compound to be as effective as misonidazole at a similar administered dose, in enhancing the action of melphalan. Misonidazole and desmethylmisonidazole also show comparable enhancement of BCNU-induced damage in the KHT tumour (31). However, this is not so in the WHF1B tumour where desmethylmisonidazole is much less effective than misonidazole in potentiating CY (28).

Various factors have been discussed that can influence the potentiating effect of nitroimidazoles on damage induced by alkylating agents. However, the nature of the cytotoxic drug itself can also greatly influence the magnitude of the potentiation as illustrated by the data of Mulcahy, Siemann and co-workers (31,32).

These authors examined the effect of misonidazole on the action of the nitrosoureas BCNU and CCNU against the KHT tumour and it was shown that misonidazole plus CCNU provided substantially greater regrowth delay than any combination of BCNU and misonidazole. The wide variation in responses observed with different tumours, timing and drug dose, suggest that for a given drug combination there will be a wide spectrum of effectiveness. Further any therapeutic value will depend upon the enhancement of tumour response being greater than any enhancement of normal tissue damage.

Evidence in favour of a differential tumour effect does exist, although more information is required. Table 3 includes some qualitative data indicating that a differential effect occurs in some experimental tumour systems.

With the exception of the results of Tannock (25,26), the data generally support the existence of a therapeutic gain. It has been pointed out, however, by Law et al. (29) that the conclusions of Tannock regarding the combination of misonidazole plus CY <u>may</u> be misleading since the comparison between normal tissue damage and tumour response was made by comparing the effects of 200 mg/kg CY alone with 75 kg/kg CY plus 1 mg/g misonidazole. The data of Law et al. (29) showed that in their system, the greatest enhancements of tumour response by misonidazole occur only at doses of CY $\leqslant$75 mg/kg. At higher doses of CY, they found that the enhancement became smaller and at doses $\geqslant$150 mg/kg, there was no longer any therapeutic gain.

MECHANISMS OF CHEMOPOTENTIATION

Various mechanisms have been proposed for the chemopotentiation phenomenon. These may be briefly reviewed.

<u>Selective killing by nitroimidazoles of drug-resistant hypoxic cells</u>

Studies involving the <u>in vitro</u> assay of tumour response to cytotoxic drugs <u>in vivo</u> generally do not indicate a large degree of sparing of hypoxic cells (2,3). Nitro compounds could selectively kill these cells but there is no other evidence to suggest that the observed enhancements are due to an additive response. Indeed, in some circumstances the nitro compounds alone cause no measurable cell killing, but nevertheless significant potentiation is still observed (see, for example Siemann (32)).

<u>Changes in drug pharmaco-kinetics</u>

It is possible that the administration of a nitro compound, misonidazole in particular, may cause a change in the pharmacokinetic behaviour of the second drug. Support for this comes from the work of Clutterbuck et al. (33) who showed increased plasma levels of melphalan when this drug was given to mice together with 1 mg/g misonidazole.

Table 3. Effects of misonidazole on normal tissue damage caused by cytotoxic drugs *in vivo*.

Cytotoxic agent	Assay of normal tissue damage	Therapeutic Ratio	Reference
CY	Numbers of deaths in treated population, weight loss, white cells	-	26
CY	LD_{50}, weight loss, white cells	+	30
CY	LD_{50}, white cells, bladder epithelium	+	28
CY	Bone marrow stem cells, white cells	+	29
CY	LD_{10}, bone marrow stem cells	+	24
Melphalan	Bone marrow stem cells, intestinal crypt stem cells	+	23
Melphalan	LD_{10}, bone marrow stem cells	+	24
CCNU	LD_{50}	+	32
BCNU	Number of deaths in treated population, weight loss, white cells	±	26
Methotrexate	Number of deaths in treated population, weight loss, white cells	-	25

+ = drug combination results in therapeutic gain
- = greater effect on normal tissues than on tumour response

Further, Tannock (26) reported that misonidazole prolonged the presence of active metabolites in the serum of mice. These effects may be due, in part, to the high doses of misonidazole leading to a reduction in the body temperature of mice (37). Although this may account for the small enhancements of normal tissue damage sometimes observed (23-26,28-30,32) it is unlikely to explain the differential effect on tumour cells compared with normal cells. Evidence suggesting that a change in pharmacokinetics may be no more than a partial explanation, comes from the observation that the 2-nitroimidazole SR 2508, which does not cause any temperature decrease in mice, sensitizes the RIF 1 tumour to CY to a similar degree as misonidazole (29). Further, the shape of dose response curves for misonidazole (see, for example Fig. 1) indicates misonidazole can still be active at low doses, where one would expect that any effect of misonidazole on the pharmacokinetics of the alkylating agent would be much reduced or even absent.

Interference with the repair of PLD

Evidence that misonidazole may suppress or interfere with the capacity of the cell to repair potentially-lethal damage (PLD) comes from the work of Law et al. (29). These authors used the clonogenic assay for cell survival to show that when tumour-bearing mice were treated with CY alone, the extent of cell-killing decreased as the interval between drug-treatment and excision of the tumour for assay was increased from 6-24 hours. This was attributed to the repair of CY-induced PLD. No such time-dependent decrease in the extent of cell-killing was observed for mice given a combined treatment of misonidazole and cyclophosphamide. However, not all tumours exhibit repair of PLD and this includes the Lewis lung tumour where misonidazole has been shown to produce large enhancements of cytotoxic damage (23,27).

Hypoxia-mediated potentiation

Another mechanism which may play a part in the observed *in vivo* responses is strongly supported by some *in vitro* experiments which implicate a hypoxia-mediated phenomenon (38,39). It was noted that the shape of the survival curve for Chinese hamster cells exposed to misonidazole under hypoxic conditions showed an initial lag period after which survival decreased exponentially as a function of contact time with drug. It was suggested that this lag period in the survival curve might be due to the time required for the build-up of some toxic metabolite of misonidazole and/or the accumulation of sub-lethal damage (SLD). Experiments were carried out in which cells were exposed to misonidazole for a time that was not sufficient to cause much cell-kill. These treated cells were then subsequently exposed to melphalan or other cytotoxic agents. The cells showed much greater sensitivity to the second cytotoxic treatment compared with cells not receiving the pre-treatment with misonidazole. Data showing the increased cell killing that can occur are given in Table 4. For both exponentially-growing and plateau-phase cells, pre-treatment with misonidazole greatly increases the cell killing effect of melphalan, even though treatment with misonidazole alone has relatively little effect. Other nitro-compounds, such as metronidazole and nitrofurazone, also show this pre-incubation effect. The potentiation is likely to be redox-related since for a given degree of potentiation the required exposure (i.e. concentration x time) to the nitro-compound decreases with increasing electron-affinity of the compound.

THE ROLE OF SULPHYDRYL COMPOUNDS

One of the effects of misonidazole pre-treatment may be to oxidise glutathione (GSH) and other non-protein sulphydryls (NPSH). It is known that cells lose endogenous NPSH following hypoxic treatment with misonidazole (40) and that addition of exogenous sulphydryls protects against this depletion. Regeneration of GSH after misonidazole treatment can occur and does so over a period of several hours (Bump and Brown, personal communication). Our own experiments show that the sub-lethal damage caused by misonidazole under hypoxic conditions is repairable in cells harvested from exponentially-growing cultures and that the time scale of this repair is 2 to 4 hours (39). Since there is good evidence to suggest that the presence of sulphydryls can also affect the response of cells to alkylating damage (41) it is possible that intracellular thiol levels can influence not only the *in vitro* pre-incubation effect but also influence the chemopotentiating effect observed in experimental tumours *in vivo*.

Table 4. The effect of pre-treating Chinese hamster cells with 5 mM misonidazole for 1½ hours under hypoxic conditions at 37°C prior to exposure to melphalan in air (taken from ref. 39).

	Surviving Fraction		
	Cells pre-treated with misonidazole alone	1 μg/ml melphalan for 1 hour	Misonidazole pre-treatment followed by melphalan
Exponentially growing cells	0.8	0.4	0.6×10^{-2}
Plateau phase cells	0.9	0.8×10^{-1}	1.3×10^{-3}

REFERENCES

1. W. U. Shipley, J. A. Stanley and G. G. Steel, Tumour size dependency on the radiation response of the Lewis Lung carcinoma, Cancer Res. 35:2488 (1975).
2. B. Dixon, J. V. Moore and H. Speakman, Radiobiological hypoxia of a transplanted rat tumour and the effect of treatment with cyclophosphamide, Eur. J. Cancer 14:1383 (1978).
3. R. P. Hill and J. A. Stanley, The response of hypoxic B16 melanoma cells to in vivo treatment with chemotherapeutic agents, Cancer Res. 35:1147 (1975).
4. J. W. Harris and D. C. Shrieve, Effects of adriamycin and X-rays on euoxic and hypoxic EMT6 cells in vitro, Int. J. Radiat. Oncol. Biol. Phys 5:1245 (1979).
5. B. A. Teicher, J. S. Lazo and A. C. Sartorelli, Classification of antineoplastic agents by their selective toxicities towards oxygenated and hypoxic cells, Cancer Res. 41:73 (1981).
6. I. Tannock and P. Guttman, Response of Chinese hamster ovary cells to anti-cancer drugs under aerobic and hypoxic conditions, Br. J. Cancer 43:245 (1981).
7. E. Smith, I. J. Stratford and G. E. Adams, Cytotoxicity of Adriamycin on aerobic and hypoxic Chinese hamster V79 cells in vitro, Br. J. Cancer 41:568 (1980).
8. E. Smith, The influence of hypoxia on the cytotoxic response of Chinese hamster cells, Ph.D. Thesis, University of London (1981).
9. W. M. C. Martin and N. J. McNally, Cytotoxicity of Adriamycin to tumour cells in vivo and in vitro, Br. J. Cancer 42: 881 (1980).
10. L. Roizin-Towle and E. J. Hall, Studies with Bleomycin and Misonidazole on aerated and hypoxic cells, Br. J. Cancer 37:254 (1978).
11. G. E. Adams, K. B. Dawson and I. J. Stratford, Electron-affinic radiation sensitizers for hypoxic cells: prospects and limitations with present and future drugs, in: Proc. Int. Meeting for Radio-Oncology, May 1978, Vienna, Georg Thiéme, Stuttgart:84 (1980).
12. I. J. Stratford, C. Williamson and G. E. Adams, Combination studies with misonidazole and a cis-Platinum complex: Cytotoxicity and Radiosensitization in vitro, Br. J. Cancer 41:517 (1980).
13. E. J. Hall and L. Roizin-Towle, Hypoxic sensitizers: Radiobiological studies at the cellular level, Radiology 117: 453 (1975).
14. J. K. Mohindra and A. M. Rauth, Increased cell killing by metronidazole and nitrofurazone of hypoxic compared to aerobic mammalian cells, Cancer Res. 36:930 (1976).

15. B. A. Moore, B. Palcic and L. D. Skarsgard, Radiosensitizing and toxic effects of the 2-nitroimidazole Ro-07-0582 in hypoxic mammalian cells, Radiat. Res. 67:459 (1976).
16. I. J. Stratford and G. E. Adams, The effect of hyperthermia on differential cytotoxicity of a hypoxic cell radiosensitizer, Ro 07-0582, on mammalian cells *in vitro*, Br. J. Cancer 35:307 (1977).
17. R. M. Sutherland, Selective chemotherapy of non-cycling cells in an *in vitro* tumour model, Cancer Res. 34:3501 (1974).
18. R. Sridhar, C. Koch and R. M. Sutherland, Cytotoxicity of two nitroimidazole radiosensitizers in an *in vitro* tumour model, Int. J. Radiat. Oncol. Biol. Phys. 1:1145 (1976).
19. R. M. Sutherland, The multicellular spheroid system as a tumour model for studies of radiation sensitizers, Pharmac. Ther. 8:105 (1980).
20. J. L. Foster, P. J. Conroy, A. J. Searle and R. L. Willson, Metronidazole (Flagyl): Characterization as a cytotoxic drug specific for hypoxic cells, Br. J. Cancer 33:485 (1976).
21. J. M. Brown, Cytotoxic effects of the hypoxic cell radiosensitizer Ro-07-0582 to tumour cells *in vivo*, Radiat. Res. 72:469 (1977).
22. D. V. Ash, M. R. Smith and R. D. Bugden, Distribution of misonidazole in human tumours and normal tissues, Br. J. Cancer 39:503 (1979).
23. C. M. Rose, J. L. Millar, J. H. Peacock, T. A. Phelps and T. C. Stephens, Differential enhancement of melphalan cytotoxicity in tumour and normal tissue by misonidazole, *in*: "Radiation Sensitizers: Their use in the clinical management of cancer," Cancer Management Vol. 5, ed., L. W. Brady, Masson, New York (1980) p.250.
24. J. J. Clement, M. S. Gorman, I. Wodinsky, R. Catane and R. K. Johnson, Enhancement of anti-tumour activity of alkylating agents by the radiation sensitizer misonidazole, Cancer Res. 40:4165 (1980).
25. I. Tannock, *In vivo* interaction of anti-cancer drugs with misonidazole or metronidazole: Methotrexate, 5-fluorouracil and adriamycin, Br. J. Cancer 42:861 (1980).
26. I. Tannock, *In vivo* interaction of anti-cancer drugs with misonidazole or metronidazole: Cyclophosphamide and BCNU, Br. J. Cancer 42:871 (1980).
27. T. C. Stephens, V. D. Courtenay, J. Mills, J. H. Peacock, C. M. Rose and D. Spooner, Enhanced cell killing in Lewis Lung carcinoma and a human pancreatic-carcinoma xenograft by the combination of cytotoxic drugs and misonidazole, Br. J. Cancer 43:451 (1981).
28. W. M. C. Martin, N. J. McNally and J. de Ronde, Enhancement of the effect of cytotoxic drugs by radiosensitizers, Br. J. Cancer 43:756 (1981).

29. M. P. Law, D. G. Hirst and J. M. Brown, The enhancing effect of misonidazole on the response of the RIF 1 tumour to cyclophosphamide, Br. J. Cancer in press (1981).
30. P. R. Twentyman, Modification of tumour and host response to cyclophosphamide by misonidazole and by WR2721, Br. J. Cancer 43:745 (1981).
31. R. T. Mulcahy, D. W. Siemann and R. M. Sutherland, In vivo response of KHT sarcomas to combination chemotherapy with radiosensitizers and BCNU, Br. J. Cancer 43:93 (1981).
32. D. W. Siemann, The in vivo combination of the nitroimidazole misonidazole and the chemotherapeutic agent CCNU, Br. J. Cancer 43:367 (1981).
33. R. Clutterbuck, J. L. Millar and T. McElwain, Misonidazole enhancement of the action of BCNU and melphalan against human melanoma xenografts, Cancer Clin. Trials 4: in press (1981).
34. G. E. Adams, C. Clarke, K. B. Dawson, P. W. Sheldon and I. J. Stratford, Nitroimidazoles as hypoxic cell radiation sensitizers and cytotoxic agents, Proc. 2nd Int. Symposium: Biological Bases and Clinical Implications of Tumour Resistance, Rome, Italy, September 21-24 (1980).
35. N. J. McNally, J. Denekamp, P. W. Sheldon, I. R. Flockhart and F. A. Stewart, Radiosensitization by misonidazole (Ro-07-0582): The importance of timing and tumour concentrations, Radiat. Res. 73:568 (1978).
36. J. M. Brown and N. Y. Yu, The optimum time for irradiation relative to tumour concentration of hypoxic cell sensitizers, Br. J. Radiol. 53:915 (1980).
37. C. J. Gomer and R. J. Johnson, The relationship between misonidazole toxicity and core temperature in C3H mice, Radiat. Res. 78:329 (1979).
38. I. J. Stratford, G. E. Adams, M. R. Horsman, S. Kandaiya, S. Rajaratnam, E. Smith and C. Williamson, The interaction of misonidazole with radiation, chemotherapeutic agents, or heat: A preliminary report, Cancer Clinical Trials 3:231 (1980).
39. E. Smith, I. J. Stratford and G. E. Adams, Misonidazole enhancement of the cytotoxic action of melphalan in vitro, Br. J. Cancer, submitted (1981).
40. M. E. Varnes, J. E. Biaglow, C. J. Koch and E. J. Hall, Depletion of nonprotein thiols of hypoxic cells by misonidazole and metronidazole, in: "Radiation Sensitizers: Their use in the clinical management of cancer," Cancer Management vol. 5, ed. L. W. Brady, Masson, New York, (1980) p. 121.
41. T. A. Conners, Protection against the toxicity of alkylating agents by thiols: The mechanism of protection and its relevance to cancer chemotherapy - a review, Eur. J. Cancer 2:293 (1966).

METRONIDAZOLE AND MISONIDAZOLE IN CANCER CHEMOTHERAPY : EFFECT OF CELL DENSITY ON DRUG ACTIVITY

J. S. Mahood and R. L. Willson

Biochemistry Department
Brunel University
Uxbridge, Middlesex, U.K.

INTRODUCTION

Metronidazole (Flagyl) has been widely used for many years for the treatment of infections by anaerobic microorganisms (1).

The original choice of the drug for laboratory screening as a hypoxic radiosensitiser (2) was based on the fact that it contained a "nitro" group and was therefore likely to sensitise by interfering with free radical reactions according to the 'electron affinity' and 'fixation' models previously proposed (3-5). Although the drug was known to be selectively cytotoxic to anaerobic microorganisms (6) hope that it might be of use as a cytotoxic agent for hypoxic cells was soon dampened: in radiosensitising studies the drug showed little toxicity at concentrations up to 10 mM in the absence of radiation, towards V.79 Chinese hamster cells exposed for 2 hours at room temperature under nitrogen 5% CO_2 on glass dishes in minimal medium supplemented with 15% serum (7). Hopes were raised again however when in the absence of radiation it was found that an increased rate of cell loss from tumours occurred when animals were given metronidazole (8) and in spheroid culture non-cycling mammalian cells were selectively killed (9).

Since this time the toxic action of metronidazole and the related nitroimidazole, misonidazole, towards hypoxic cells both *in vitro* and *in vivo* has been demonstrated in many laboratories (10-21). Evidence reported to date indicates that misonidazole is the more potent cytotoxic agent towards hypoxic mammalian cells *in vitro*. *In vivo*, however, studies in which tumour-bearing mice were given a course of each drug followed by radiation treatment after an interval sufficient to allow the drug to metabolise, indicate that metronidazole is the more effective (11,14). This higher activity of metronidazole has also been observed in several anaerobic bacterial and protozoal systems as well as with the bacteria Serratia marcescens, E. coli B/r and E. coli WP.2 incubated under hypoxic conditions (6,18).

We now present evidence that the cytotoxicity of metronidazole towards hypoxic mammalian cells is dependent on cell density. We have also shown that the cell killing at high cell density can be duplicated at low density by addition of 20 mM sodium lactate to the medium. Under these conditions the activity of metronidazole approximates to that of misonidazole.

MATERIALS AND METHODS

CHO-K1 cells were grown in Ham F.12 medium (Flow Laboratories) supplemented with 10% foetal bovine serum at 37°C in an atmosphere of 5% CO_2: 95% air. Viable cells were determined by colony counting after serial dilution and growth for 7-10 days on 5 cm dishes. Cells were suspended at the appropriate density at 37°C in medium containing 7% serum in 50 ml Bellco spinner flasks and gently stirred at 120 r.p.m. with 95% N_2, 5% CO_2 (1 l.per min) flowing over the surface of the medium. Metronidazole was kindly supplied by May and Baker Ltd. and misonidazole by Professor G. E. Adams and Roche Products Ltd. Lactic acid (Sigma L+) was titrated to pH = 7.4 with sodium hydroxide before use.

RESULTS

The effect of metronidazole (5 mM on the viability of hypoxic cells incubated at low and high cell densities is shown (Fig. 1). The absence of significant killing at the density of 2 x 10^4 cells per ml is in accord with the results of other workers (12,21). At a density of 5 x 10^6 cells per ml little killing occurred over 2 hours: significant killing then took place.

The effect of metronidazole and misonidazole on the viability of hypoxic cells incubated at a density of 2 x 10^4 cells per ml in the absence and presence of 20 mM sodium lactate is shown (Fig. 2). In the absence of lactate misonidazole (5 mM) showed little toxicity up to 6 hours: significant killing then followed. In the presence of lactate, however, the activity of both drugs increased considerably, the cell viability after 10 hours falling from 55.4 to 0.16% in the case of metronidazole and from 7.5 to 0.09% with misonidazole. In both instances the apparent drug resistance at shorter times of incubation was removed almost entirely. Sodium lactate (20 mM) alone showed little toxicity over 10 hours of incubation.

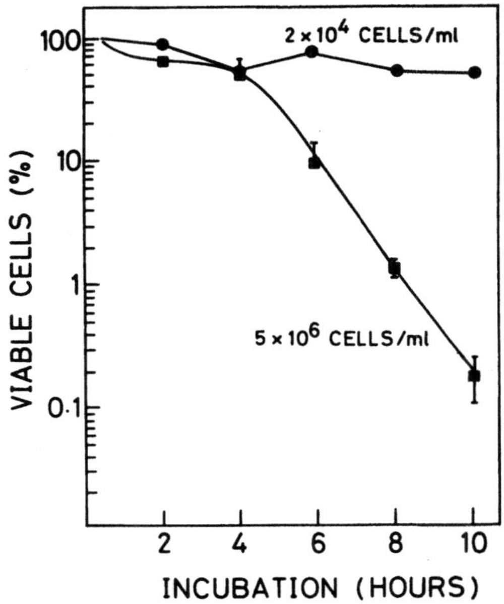

Fig. 1. Percentage viability of hypoxic CHO cells incubated at different initial cell densities with 5 mM metronidazole.

DISCUSSION

The present results clearly indicate that in assessing drug activity the cell density must be taken into account. It was initially thought that the increased killing at high cell density was due to the increased probability of possible traces of oxygen being removed by metabolism: oxygen is known to inhibit the cytotoxic action of the drug (11,13,17,18). However, in view of the results with lactate it is thought that the increased cell killing is more likely to be associated with the greater production of this metabolite.

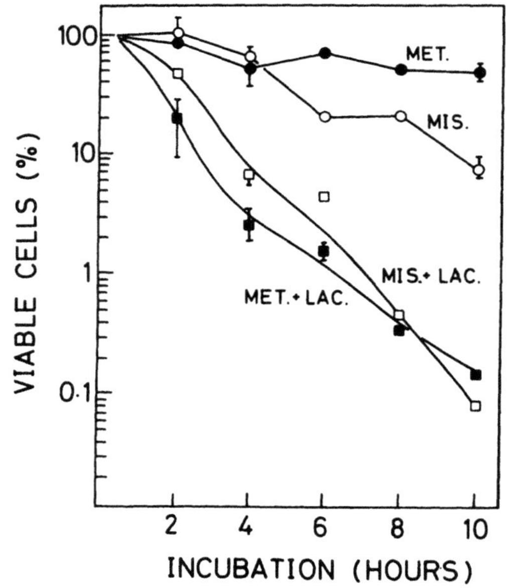

Fig. 2. Percentage viability of hypoxic CHO cells incubated with 5 mM metronidazole or misonidazole in the absence and presence of 20 mM sodium lactate (initial cell density 2×10^4 cells/ml).

This dramatic effect of lactate (Fig. 2) must be of relevance to the *in vivo* situation. It as seems likely, solid tumours contain relatively high lactate concentrations, metronidazole may be at least as effective as misonidazole in combination therapy.

The possibility that the observed increase in activity is due solely to a fall in pH associated with the production of lactic acid rather than lactate itself is considered unlikely in view of the buffering capacity of the system. We feel it is more likely that the 'lactate effect' is due to an increased activation of the drugs and/or an increased sensitivity of the cells to them with the lactate-pyruvate redox couple catalysed by the enzyme lactate dehydrogenase (LDH) being somehow involved.

$$\text{lactate} + NAD^{+} \overset{LDH}{\rightleftharpoons} \text{pyruvate} + NADH + H^{+}$$

Through this equilibrium the reduction potential of the cell may be increased leading to an increased activation of the drug. The presence of excess lactate is also likely to lead to an increase in the levels of the LDH-NADH complex known to be sensitive to free radical attack (23). The free radical $RNO_2^{\cdot -}$, the nitroimidazole analogue of the oxygen radical, $O_2^{\cdot -}$ has been suggested as a toxic intermediate. One of the few oxidative reactions of $O_2^{\cdot -}$ documented is that with the LDH-NADH complex. The following chain reaction occurs :

$$LDH + NADH \rightleftharpoons LDH - NADH$$
$$O_2^{\cdot -} + LDH - NADH \rightleftharpoons HO_2^{-} + LDH + NAD^{\cdot}$$
$$LDH - NAD^{\cdot} + O_2 \rightleftharpoons LDH - NAD^{+} - O_2^{\cdot -}$$
$$\text{and} \quad HO_2^{-} + H_2O \rightleftharpoons H_2O_2 + OH^{-}$$

An analogous chain reaction with RNO_2 replacing O_2 can be envisaged with the formation of the nitroso compound instead of hydrogen peroxide.

$$RNO_2^{\cdot -} + LDH - NADH \rightleftharpoons RNO_2H^{-} + LDH - NAD^{\cdot}$$
$$LDH - NAD^{\cdot} + RNO_2 \rightleftharpoons LDH - NAD^{+} + RNO_2^{\cdot -}$$
$$RNO_2H^{-} \rightleftharpoons RNO + OH^{-}$$

In the presence of acid, synergism may occur since the protonated electron adduct $RNO_2H^{\cdot}$ is likely to be a stronger oxidant than $RNO_2^{\cdot -}$. Pulse radiolysis and related stationary-state studies are currently being used to investigate the likelihood of these reactions.

SUMMARY

The cytocidal action of metronidazole on hypoxic mammalian cells *in vitro* has been found to be strongly dependent on cell density. The activity at high cell density can be duplicated at low density by the addition of 20 mM sodium lactate to the incubating medium. The effect of lactate on the cytotoxicity of misonidazole is less pronounced. The results suggest that the present preferential use of misonidazole in cancer therapy may need to be reassessed.

ACKNOWLEDGEMENTS

Financial support from the Cancer Research Campaign is gratefully acknowledged. J.S.M. is holder of an S.R.C. C.A.S.E. Award in conjunction with May and Baker Ltd.

1. E. J. Baines, Metronidazole: sixteen years of developments: metronidazole, in: Excerpta Medica, S. M. Finegold ed. p.61, and refs cited (1977).
2. J. L. Foster and R. L. Willson, Radiosensitization of anoxic cells by metronidazole, Br. J. Radiol., 46:234 (1973).
3. P. T. Emmerson and P. Howard Flanders, Sensitization of anoxic bacteria to X-rays by organic nitroxide free radicals, Radiat. Res., 26:54 (1965).
4. G. E. Adams and D. L. Dewey, Hydrated electrons and radiobiological sensitization, Biochem. Biophys. Res. Comm., 12:473 (1963).
5. R. L. Willson and P. T. Emmerson, Reaction of triacetoneanine-N-oxyl with radiation-induced radicals from DNA and from deoxyribonucleotides in aqueous solution, *in* : "Radioprotection and Sensitization", H. L. Moroson and M. Quintilian ed., Taylor and Francis Ltd., London (1970).
6. H. N. Prince, E. Grunberg, E. Titsworth and W. F. Delorenzo, Effects of 1-(2-nitro-1-imidazole)-3-methoxy-2-propanol and 2-methyl-5-nitroimidazole-1-ethanol against anaerobic and aerobic bacteria and protozoa, Applied Microbiol., 18: 728 (1969).
7. J. C. Asquith, J. L. Foster and R. L. Willson, Metronidazole (Flagyl). A radiosensitizer of hypoxic cells, Br. J. Radiology, 47:474 (1974).
8. A. C. Begg, P. W. Sheldon and J. L. Foster, Demonstration of hypoxic cells in solid tumours by metronidazole, Br. J. Radiol. 47:399 (1974).
9. R. M. Sutherland, Selective chemotherapy of non-cycling cells in an *in vitro* tumour model, Cancer Res., 34:3501 (1974).
10. J. L. Foster and R. L. Willson, Metronidazole (Flagyl) in cancer radiotherapy, Chemotherapy, 7:215 (1976).
11. J. L. Foster, P. J. Conroy, A. J. Searle and R. L. Willson, Metronidazole (Flagyl): characterization as a cytotoxic drug specific for hypoxic tumour cells. Br. J. Cancer, 33:485 (1976).
12. J. K. Mohindra and A. M. Rauth, Increased cell killing by metronidazole and nitrofurazone of hypoxic compared to aerobic mammalian cells, Cancer Res., 36:930 (1976).
13. R. Sridhar, C. Koch and R. Sutherland, Cytotoxicity of two nitroimidazole radiosensitizers in an *in vitro* tumour model. Int. J. Radiat. Oncol. Biol. Phys., 1:1149(1976).

14. J. L. Foster, Differential cytotoxic effects of metronidazole and other nitro heterocyclic drugs against hypoxic tumour cells, Int. J. Radiat. Oncol. Biol. Phys., 2:Suppl.III:153 (1977).
15. E. J. Hall, M. Astor, C. Geard and J. Biaglow, Cytotoxicity of Ro-07-0582: enhancement by hyperthermia and protection by cysteamine, Br. J. Cancer, 35:809 (1977).
16. E. J. Hall, R. Miller, M. Astor and F. Rini, The nitroimidazoles as radiosensitizers and cytotoxic agents. Br. J. Cancer, 37:Suppl.III:120 (1978).
17. Y. C. Taylor and A. M. Rauth, Differences in the toxicity and metabolism of the 2-nitroimidazole misonidazole (Ro-07-0582) in HeLa and Chinese hamster ovary cells. Cancer Res., 38:2745 (1978).
18. S. H. Basaga, J. R. Dunlop, A. J. F. Searle and R. L. Willson, Metronidazole (Flagyl) and Misonidazole (Ro-07-0582): reduction by facultative anaerobes and cytotoxic action on hypoxic bacteria and mammalian cells in vivo. Br. J. Cancer, 37:Suppl.III:132 (1978).
19. I. J. Stratford, Split dose cytotoxic experiments with misonidazole. Br. J. Cancer, 38:130 (1978).
20. E. O. Pettersen, Toxic and radiosensitizing effect of the 2-nitroimidazole Misonidazole (Ro-07-0582) on murine CFU in vivo. Br. J. Cancer 37:Suppl.III:107 (1978).
21. G. E. Adams, I. J. Stratford, R. G. Wallace, P. Wardman and M. E. Watts, Toxicity of nitrocompounds towards hypoxic mammalian cells in vitro: dependence on reduction potential. J. Nat. Cancer Inst., 64:555 (1980).
22. D. W. Whillans and A. M. Rauth, An analysis of changes in oxygen tension in stirred cellular suspensions under conditions of radiolytic and cellular consumption. Cancer Clin. Trials, 3:63 (1980).
23. B. H. J. Bielski and P. C. Chan, Kinetic study by pulse radiolysis of the lactate dehydrogenase-catalyzed chain oxidation of nicotinamide adenine dinucleotide by HO_2 and O_2^- radicals. J. Biol. Chem., 250:318 (1974).

METABOLISM OF NITROIMIDAZOLES

D.E. Schwartz and W. Hofheinz

Research Departments, F. Hoffmann-La Roche & Co.Ltd.
Basle, Switzerland

The development in recent years of nitroimidazoles as chemotherapeutic agents has been extensive. The interest in this class of compounds apparently was raised in 1955 by the elucidation of the structure of the antibiotic Azomycin by Nakamura(1) and the discovery of the in vivo activity of 5-nitro-imidazoles against Trichomonas vaginalis(2). Since then an ever growing number of 2- and 5-nitro-imidazoles have been synthesized.

Azomycin

Due to major difficulties met in the synthesis of the 2-nitro-imidazoles developments have occurred mainly in the group of 5-nitro-imidazoles. Among the most interesting compounds introduced on to the market or into clinical evaluation are the following 5-nitro-imidazoles: dimetridazole, metronidazole, ronidazole, tinidazole, ornidazole, ipronidazole, nimorazole; the 2-nitro-imidazoles : benznidazole and misonidazole.

The purpose of our presentation is to briefly review present knowledge of their metabolism. Dimetridazole (=EMTRYL) and metronidazole (=FLAGYL, CLONT) both synthesized by Rhône Poulenc, were the first compounds of this group to be investigated as to their fate in the living organism.

Dimetridazole, found to be active against a protozoan flagellate, the agent of histomoniasis in turkeys, was investigated in this species. Using the drug labelled with ^{14}C at position 2 both in the ring and in the methyl group(3,4) it was found that the main metabolic pathway involved oxidation of the 2-methyl to 2-hydroxymethyl group, which in turn was conjugated to sulphuric or glucuronic acid or further oxidised to the corresponding carboxylic acid. 90% of the excreted drug-related material consisted of metabolites which

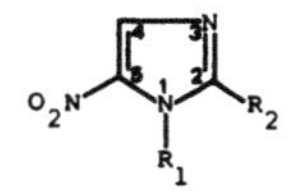

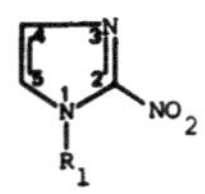

R_1	R_2	Generic name (trade name)	R_1	Generic name (trade name)
$-CH_3$	$-CH_3$	Dimetridazole (EMTRYL)	$-CH_2-CO-NH-CH_2-C_6H_5$	Benznidazole (RADANIL)
$-CH_3$	$-CH_2-O-C(=O)-NH_2$	Ronidazole (RONIDAZOLE)	$-CH_2-CHOH-CH_2-OCH_3$	Misonidazole
$-CH_3$	$-CH(CH_3)_2$	Ipronidazole (IPROPAN)		
$-CH_2-CH_2-OH$	$-CH_3$	Metronidazole (FLAGYL)		
$-(CH_2)_2-SO_2-CH_2-CH_3$	$-CH_3$	Tinidazole (FASIGYN)		
$-CH_2-CHOH-CH_2Cl$	$-CH_3$	Ornidazole (TIBERAL)		
$-CH_2-CH_2-N\bigcirc O$	$-H$	Nimorazole (NAXOGIN)		

had retained the nitro-imidazole structure. No evidence of a reduction of the nitrogroup could be detected. The metabolic scheme, although incomplete, may be summarized as follows :

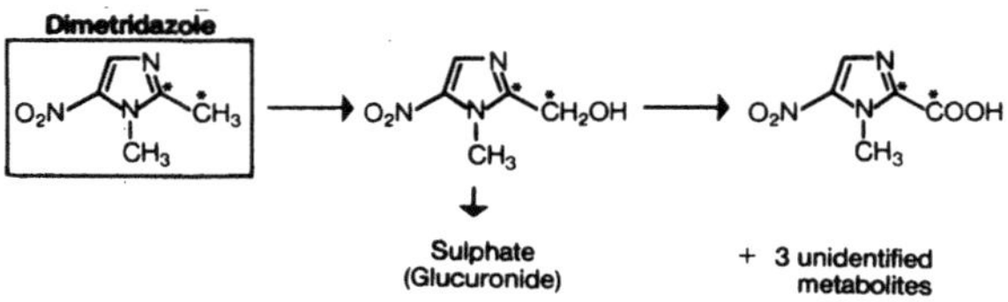

A large number of publications by several groups of investigators have dealt with the metabolism of Metronidazole. Among others Stambaugh et al(5) studied the fate of the drug in humans with non-labelled material, whereas Ings et al(6) conducted similar studies in rats with metronidazole ^{14}C-labelled at position 2 in the ring. From their data a general scheme, common to both species, may be deduced :

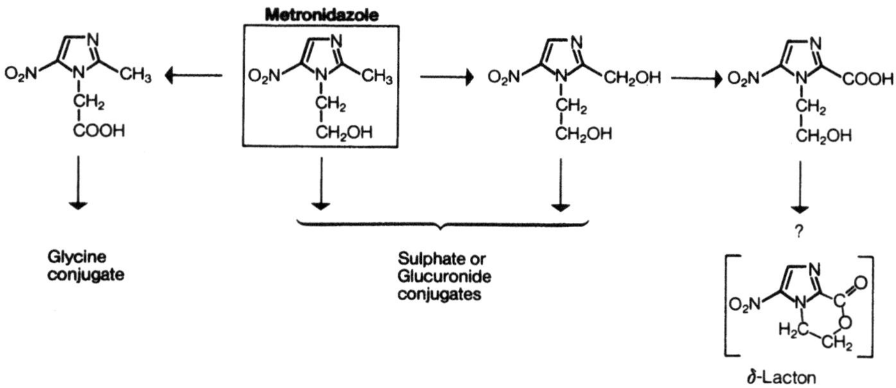

No evidence was found for the formation of the amino analogue of metronidazole as a metabolite(5,6). However, because of the known instability of 5-amino-imidazoles, the authors(5) do not exclude that such a metabolite may have undergone rapid degradation and thus escaped identification.

More recently Koch et al isolated small amounts of N(2-hydroxy-ethyl)-oxamic acid(7) and acetamid(8) from the urine of rats given metronidazole orally, indicating cleavage of the ring :

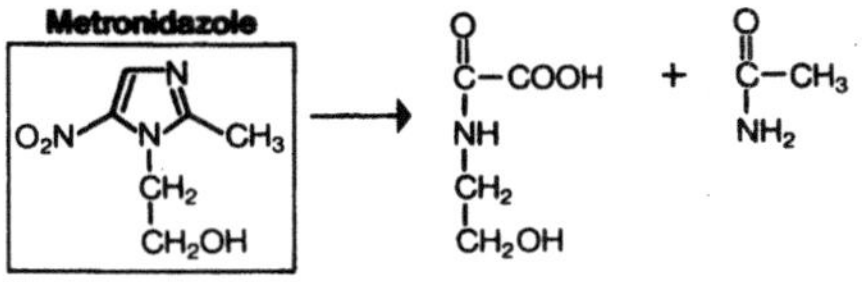

Ronidazole, a 5-nitro-imidazole, has revealed useful antiparasitic activity in the treatment of enterohepatisis of turkeys. Rosenblum et al(9,10) conducted extensive metabolic studies of the drug in this species. They used ^{14}C-Ronidazole labelled either in the N-methyl group (= I) or at position 2 in the ring (= II) :

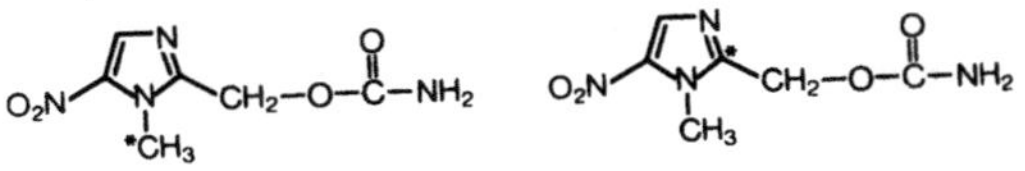

With each of the labelled compounds 80% of the administered radioactivity was excreted with the urine and the feces, while approximately 1-2% were slowly exhaled as $^{14}CO_2$. The following labelled metabolites were detected and identified by TL-chromatography or reverse isotope dilution technique :

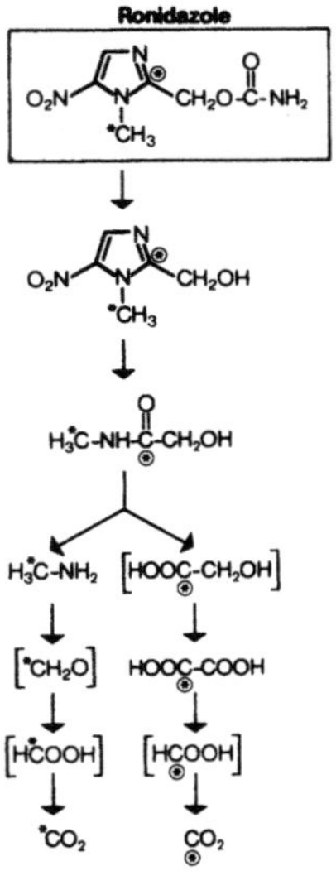

Structures in square brackets represent postulated intermediates which have not been identified.

On the basis of the structures isolated using the two labelled preparations three possible types of metabolic cleavage of the imidazole ring were considered :

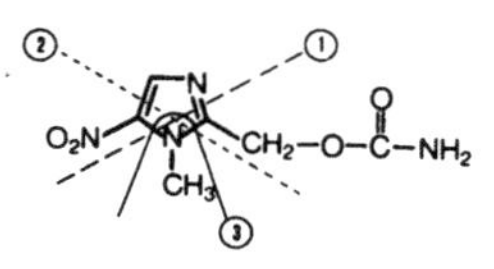

Because of the large consumption of turkey meat, particular emphasis was further given to tissues. Radioactivity was found to be associated with the C2 pool and with a number of intermediates of the glutamic, aspartic and citric acid cycles. The authors therefore concluded that ^{14}C-containing fragments of the drug, identical to endogenous compounds, were incorporated into body constituents(8). Recently this has been further substantiated(11,12). The metabolism of Tinidazole was studied in the rat and in the dog by Wood et al(13) using ^{35}S-labelled drug. The main route of excretion of radioactivity in both species was in the urine. About half of the urinary radioactivity was found to be associated with the unchanged drug. In addition four metabolites were detected, two of which were identified as the 2-hydroxymethyl derivative and its glucuronide :

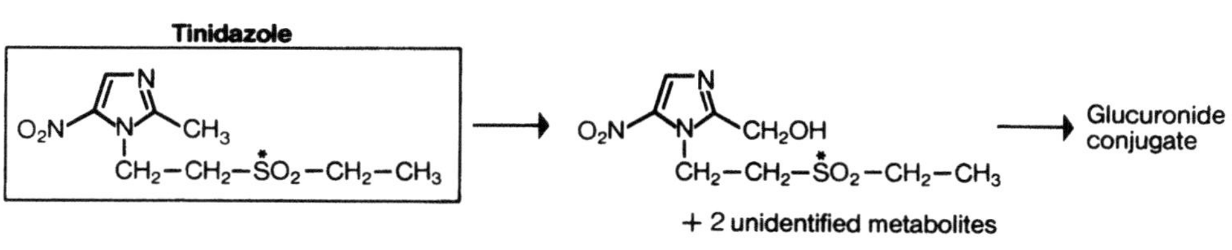

The remaining 2 unidentified metabolites "did not appear to be simple nitro-imidazoles" and were assumed to be "products of reductive biotransformation probably followed by ring opening*(13). No evidence for a reduction of the nitrogroup could be detected. (*Fig. 9).

The metabolism of Ornidazole was studied in the rat, in the dog and in man by Schwartz et al(14,15) using the drug labelled with ^{14}C at position 2 in the ring. Metabolites were extracted and isolated from urine and their structure determined by mass- and NMR-spectroscopy. The metabolic pattern of each species, although qualitatively similar, was quantitatively different. The following general metabolic scheme was proposed :

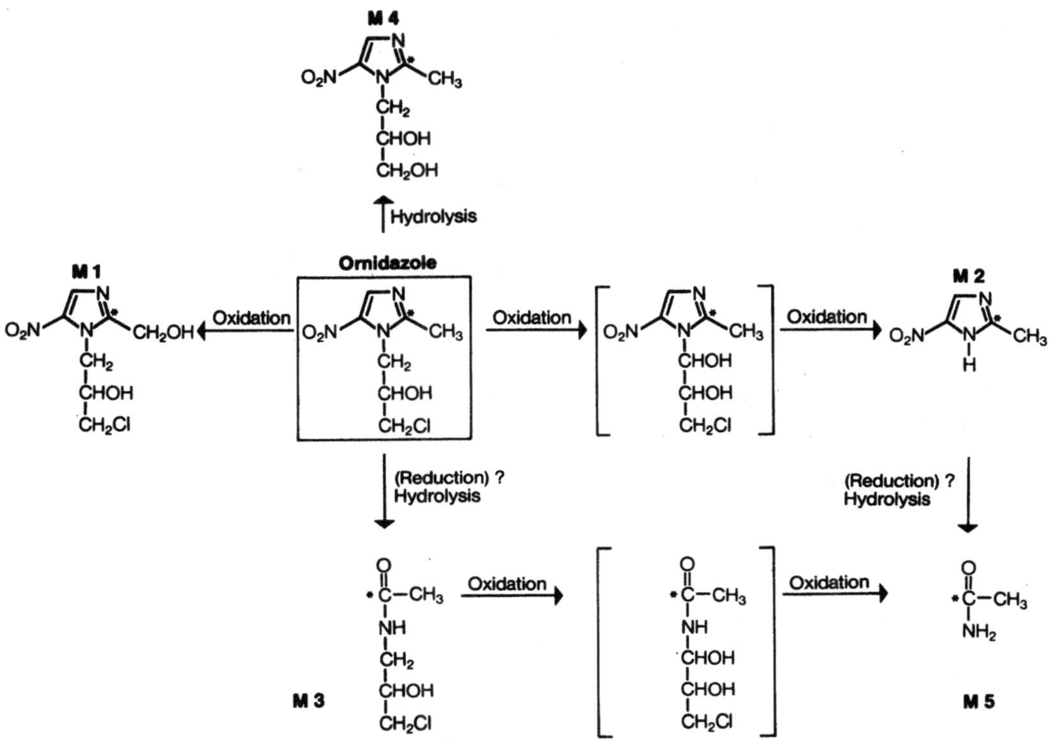

Structures in square brackets represent postulated intermediates which have not been identified.

In addition to free metabolites, conjugates (glucuronides, or sulphates) of ornidazole and of the metabolites M_1, M_2, M_3 and M_4 were detected in the urine.

Metabolites of imidazole structure with a substituent derived from the nitrogroup by reduction were not seen in any of the examined species. However the formation of the metabolites M_3 and M_5 indicated metabolic cleavage of the ring between N1/C5 and C2/N3, confirming thus one type of rupture of the ring previously postulated by Rosenblum et al(9,10). There was no indication of other scissions of the ring.

The metabolism of Flunidazole was studied in rats, dogs and humans by Zacchei et al(16) using the drug labelled with ^{14}C at position 2 in the ring. Flunidazole formed a glucuronide which prevailed in the urine of the dog and of man while it was absent in that of the rat. In addition the following two metabolites were detected in the urine in all three species. They were formed in larger proportions by the rat :

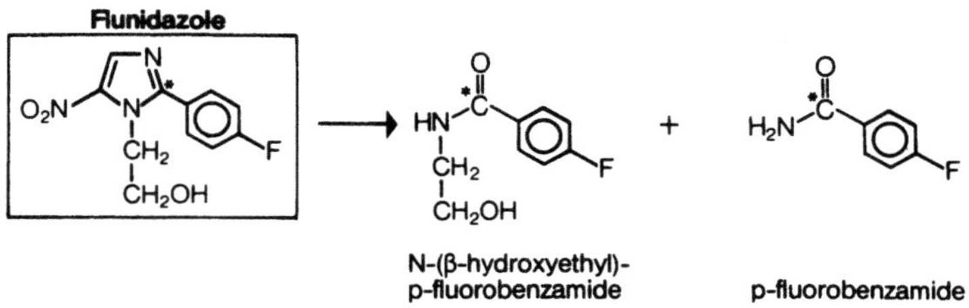

Both compounds fit a metabolic scheme similar to that suggested for Ornidazole(14) and further substantiate the cleavage of the ring between N1/C5 and C2/N3.

Two metabolites of Ipronidazole have been identified. The unchanged drug and metabolite M1 were quantitatively measured, using gas chromatography, in the tissues of turkeys(17,18) and pigs(19) given the unlabelled drug; M2 was isolated from the feces of bile cannulated rats given 2-^{14}C-labelled ipronidazole(20). Structures of both metabolites were established by mass and nuclear magnetic resonance spectroscopy. M2 may possibly be formed from M1 by addition of water.

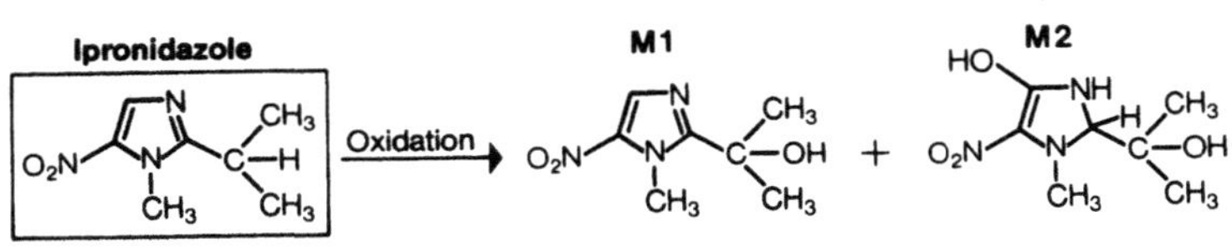

The metabolism of Nimorazole (= Nitrimidazine) was studied by Giraldi et al(21) in humans using the unlabelled drug. Two metabolites still carrying the nitrogroup were isolated from urine and their structure determined. Metabolites 1 and 2 accounted for 9% and 25% respectively of the dose.

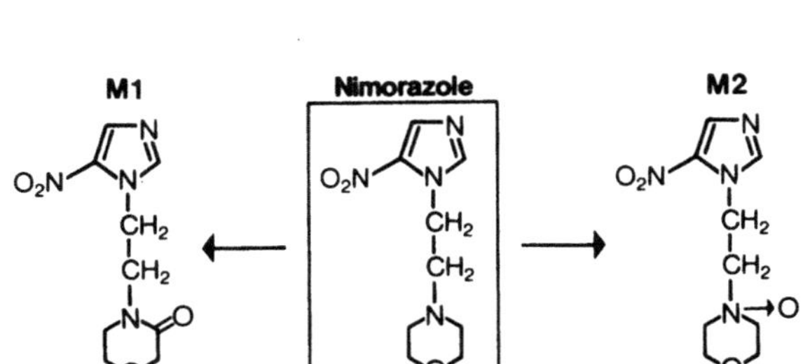

Metabolic studies of 2-nitro-imidazoles are still very scarce. Flockhart et al investigated the metabolism of Misonidazole in mice, rat, baboon and man using the drug ^{14}C-labelled at position 2 in the ring(22). In general the preferred route of excretion was in the urine. Metabolites which had retained the nitroimidazole structure accounted for more than half of this radioactivity. The following metabolites were identified :

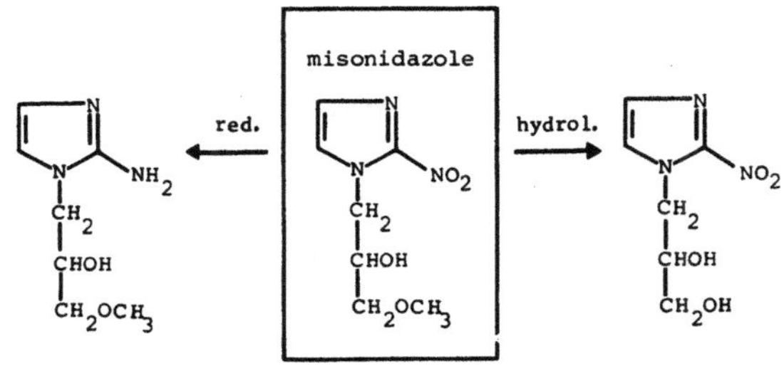

Misonidazole and desmethyl-misonidazole were also excreted in small amounts in the form of conjugates in the urine.

Metabolic studies of Benznidazole have been conducted in rats, dogs, monkeys and humans with the drug ^{14}C-labelled at position 2 in the ring. The following metabolites were identified in the urine of all 4 species(23).

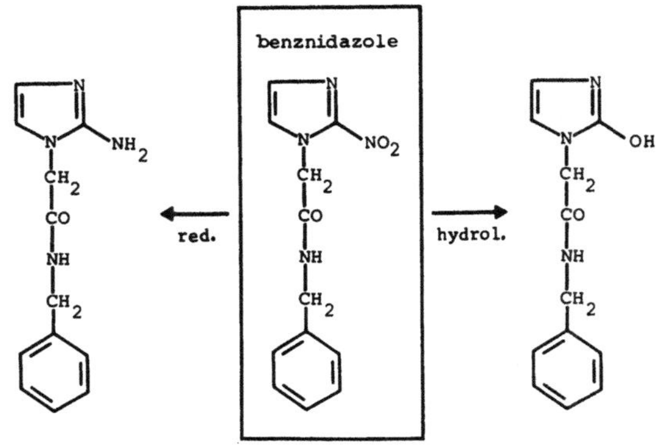

The metabolism of 1-methyl-2-nitro-5-isopropyl-imidazole, a constitutional isomer of ipronidazole, was studied in dogs using 2-^{14}C-labelled drug(24,25). In addition to the parent compound, 3 metabolites could be extracted from urine and their structures determined by infrared, mass and nuclear magnetic resonance spectroscopy. The compound indicated in square brackets represents a postulated non-identified intermediate. M2 was assumed to be also excreted in the form of conjugates in the urine. Metabolite M1 proved to be less toxic but essentially as active as the parent drug, while M2 showed a low degree of activity in vivo(26).

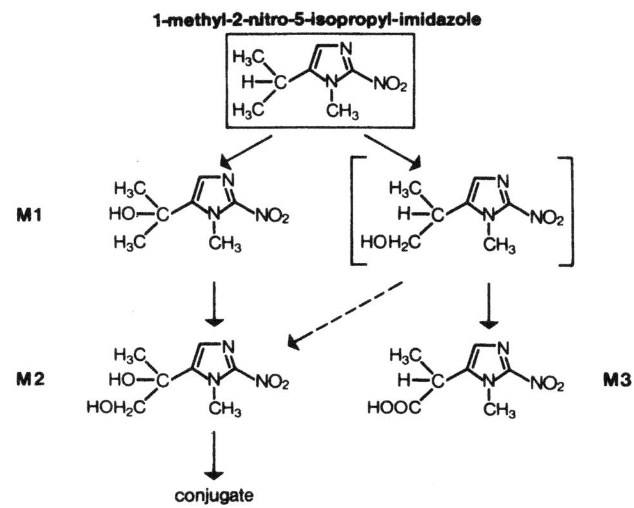

Let us now consider a particular aspect of the metabolism of nitro-imidazoles, namely that of the nitro group. The metabolic reduction of aromatic nitro compounds to their corresponding amines is well documented(27,28,29) and has been observed both in vitro (30,31) and in vivo(32,33,34). One would therefore expect nitro-imidazoles to undergo a similar type of enzymatic reduction in the organism. Indeed this has been observed, as already pointed out, in the case of 2-nitro-imidazoles.

Metabolic reduction of 5-nitroimidazoles has been proven by the fortuitous isolation by Baker et al(35) of the acetamido metabolite formed by reduction and acetylation of 1-methyl-5-nitro-2-(2'-pyrimidyl)imidazole in the rat and in man. The fact that 5-aminoimidazoles have never been detected as metabolites is not surprising in view of their well-known instability. These compounds decompose in aqueous solution at neutral pH by scission of the ring giving rise to the formation of several small fragments(36, 37,38). Such fragments have also been observed as metabolic cleavage products in vivo with ronidazole(9,10), ornidazole(14), flunidazole(16), and metronidazole(7). These observations may be taken as circumstantial evidence for the formation of aminimidazoles as transient, unstable metabolic intermediates.

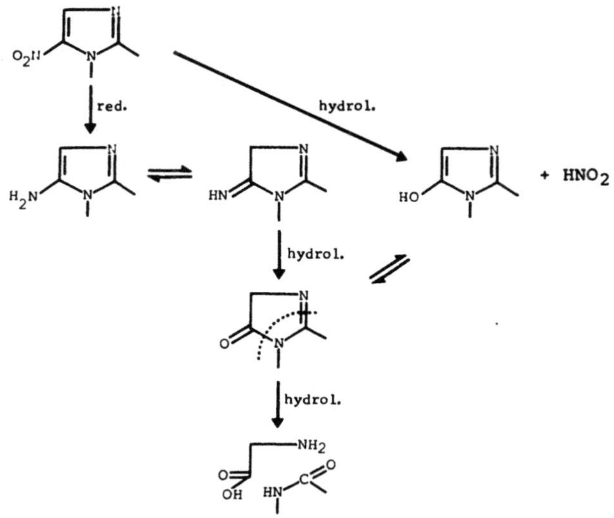

However, hydrolytic substitution of the nitro group is also a well-known chemical reaction of nitro-imidazoles. In the case of 5-nitro-imidazoles the products formed, 5-hydroxy-imidazoles, similarly to 5 amino-imidazoles, are unstable and could equally well decompose under scission of the ring. Metabolic cleavage of the imidazole ring may therefore alternatively result from an hydrolytic displacement of the nitro group with formation of nitrite ion and 5-hydroxy-imidazoles.

In the case of 2-nitro-imidazoles the corresponding 2-hydroxy-imidazoles are more stable and formation of this type of metabolite has indeed been observed with benznidazole(23).

Cleavage of the imidazole ring after reduction or hydrolysis of the nitro group by conventional chemical methods is known to occur between N1/C5, C2/N3, and N1/C2(36,37,38). The formation of metabolite M3 from ornidazole and of N-(2-hydroxyethyl)-p-fluoro-benzamide from flunidazole(14,16) conforms to this type of cleavage. A different type of scission has been observed by Koch et al(7,8) with metronidazole. It was found to occur in conventional, but not in germfree, rats and therefore appears to be mediated by the intestinal microflora. The scission products isolated require, in addition to cleavage between N1/C2 and N3/C4, a formal oxidation at C-4.

The fate of the nitro group is of particular interest also for pharmacological and toxicological reasons: Nitro-imidazoles have antiparasitic, antibacterial and radiosensitizing properties, while the corresponding amino analogues are devoid of such activities. One must therefore assume that the nitro group is essential for such activities. On the other hand nitro-imidazoles, when given over a long period of time and at high doses, disclosed carcinogenic properties in mice but rarely in other species. For all these reasons, knowledge of the metabolic fate of the drug in various species may be of prime importance. The use of 5-nitro-imidazoles labelled at position 5 in the ring may give some information on the fate of the other fragment of the molecule, but it will probably not disclose the mechanism by which the drug deploys its activitiy or toxicity. Indeed, the reduction of the nitro to the amino group involves a number of transient reactive intermediates such as various radicals, radical ions and the more stable nitroso- and hydroxylamino-derivatives. These highly reactive species may be trapped in several types of reactions and therefore escape identification.

In vitro incubation of ornidazole with rat or dog liver supernatant under aerobic conditions did not produce evidence of any metabolic degradation. In contrast the drug, when incubated under anaerobic conditions, disappeared while no metabolite with a nitro-imidazole structure could be detected(39). Similar observations were made with misonidazole (T.R. Marten, personal communication). Marked toxicity differences were observed when incubating HeLa or Chinese hamster ovary cells with misonidazole under aerobic or anaerobic conditions(40). In this respect it is of particular interest to mention that microsomal nitro-reductases are stable only in the absence of oxygen(31,41).

Further aspects of the metabolism of nitro-imidazoles remain to be clarified. Recent experiments indicate that at least some of these compounds undergo entero-hepatic recycling(42). Micro-organisms of the gastro-intestinal tract are known to possess a large number of enzymes capable of inducing an impressive variety of reactions, including the reduction of the nitro group(43). Therefore it is highly probable that microorganisms in the gut will have an influence on the metabolic pattern of nitro-imidazoles in mammals, as suggested by the experiments of Koch et al(7,8) and of Searle and Wilson(44).

SUMMARY

Depending on their structures, 2- and 5-nitro-imidazoles are metabolized to various extents in the organism. The majority of the compounds examined were found to be excreted only in minor amounts as unchanged drug in the urine. Conversion to highly polar metabolites and incorporation of small fragments of the drug molecule into normal body constituents were also observed.

Drugs carrying an alcohol group were found to form conjugates, this metabolic pathway being more important in man than in the rat. A methyl group at position 2 was a favourable point for oxidation.

Cleavage of the imidazole ring was demonstrated for a number of 5-nitro-imidazoles. This reaction may follow reduction of the nitro group. Such a reduction, likely to occur under the influence of some microsomal nitro-reductases either of the host or of some micro-organisms present in the gut, would involve a number of highly reactive intermediates. The fact that no 5-amino-imidazoles could be isolated as metabolites is not unexpected in view of their well-known chemical instability. However, the isolation of the acetamido metabolite of 1-methyl-5-nitro-2-(2'-pyrimidyl)imidazole from rat and human urine and the formation of the stable 2-amino-imidazoles as metabolites of 2-nitro-imidazoles clearly show that the nitro group can be reduced in vivo. Alternatively, ring cleavage may occur by way of hydrolytic displacement of the nitro group with formation of the hydrolytically unstable 5-hydroxy-imidazoles.

Experiments with drugs labelled at position 4 or 5 in the imidazole ring will allow further investigation of the mode of cleavage of 5-nitro-imidazoles. Studies with germ-free rats should give additional information on the relative importance of the mammalian and bacterial nitro-reductase for the rupture of the imidazole ring in vivo.

A big challenge lies ahead : that of elucidating the structures of the intermediates probably responsible for the manifold activities of these drugs.

REFERENCES

1. S. Nakamura, Structure of azomycin, a new antibiotic, Pharm.Bull. Japan 3:379 (1955).
2. C. Cosar and L. Julou, Activité de -'(hydroxy-2'-éthyl)-1-méthyl-2-nitro-5-imidazole (8.823 RP) vis-à-vis des infections expérimentales à Trichomonas vaginalis, Ann.Inst.Pasteur 96:238 (1959).
3. G.L. Law, G.P. Mansfield, D.F. Muggleton and E.W. Parnell, Dimetridazole : Absorption, excretion and metabolism in turkeys, Nature (Lond.) 197:1024 (1963).
4. H.B. Condren, R.E. Davies, C.W. Deyoe, C.R. Creger and J.R. Couch, Effects of 1,2-dimethyl-5-nitroimidazole on growth and reproduction in turkeys and its residual concentration in tissues, Poultry Sci. 41:1637 (1962)
5. J.E. Stambaugh, L.G. Feo and R.W. Manthei, The isolation and identification of the urinary metabolites of metronidazole in man, J. Pharmacol.Exp.Ther. 161:373 (1968).

6. R.M.J. Ings, J.A. McFadzean and W.E. Ormerod, The fate of metronidazole and its implications in chemotherapy, Xenobiotica 5:223 (1975).
7. R.L. Koch and P. Goldman, The anaerobic metabolism of metronidazole forms N-(2-hydroxyethyl)-oxamic acid, J. Pharmacol. Exp.Ther. 208:406 (1978).
8. R.L. Koch, E.J.T. Chrystal, B.B. Beaulieu Jr. and P. Goldman, Acetamida - a metabolite of metronidazole formed by the intestinal flora, Biochem. Pharmacol. 28:3611 (1979).
9. C. Rosenblum, N.R. Trenner and D.E. Wolf, The non-metabolite residue - a limitation to the tracer method, J. Labelled Compounds 7:225 (1970).
10. C. Rosenblum, N.R. Trenner, R.P. Buhs, C.B. Hiremath, F.R. Koniuszy and D.E. Wolf, Metabolism of ronidazole (1-methyl-5-nitroimidazole-2-ylmethyl carbamate), J.Agr. Food Chem. 20:360 (1972).
11. R.P. Buhs, A. Rosegay, T.A. Jacob, N. Allen and F.J. Wolf, Reduction of ronidazole (1-methyl-2-carbamoyl-oxymethyl-5-nitroimidazole) in vivo, Pharmacologist 21:232 Abstract No.462 (1979).
12. F.J. Wolf, T.A. Jacob, J.J. Steffen, R.F. Alvaro, M.L. Green, L.R. Chapin and R.L. Ellsworth, Ronidazole - The contributions of reductive metabolism in the formation of tissue residues, Pharmacologist 21:232, Abstract No.461 (1979).
13. B.A. Wood, D. Rycroft and A.M. Monro, The metabolism of tinidazole in the rat and dog, Xenobiotica 3:801 (1973).
14. D.E. Schwartz, J-C. Jordan, W. Vetter and G. Oesterhelt, Metabolic studies of ornidazole in the rat, in the dog and in man, Xenobiotica 9:571 (1979).
15. R. Richle, H.J. Scholer, P. Angehrn, M. Fernex, H. Hummler, F. Jeunet, K. Schärer, M. Schüpbach and D.E. Schwartz, Grundlagen der Chemotherapie von Trichomoniasis und Amoebiasis mit Ornidazol, Arzneim.-Forsch. (Drug Res) 28:612 (1978).
16. A.G. Zacchei, L. Zahniser, W.P. Freeborn and F.G. McMahon, Excretion and metabolism of 1-(2-hydroxyethyl)-2-(p-fluorophenyl)-5-nitroimidazole-2^{14}C in rat, dog and human, Fed. Proc. 29:2430 (1970).
17. J. Fellig, A. MacDonald, E. Meseck and H. Laurencot, The metabolism of ipronidazole in the turkey, Poultry Sci. 48:1806 (1969).
18. A. MacDonald, G. Chen, M. Kaykaty and J. Fellig, Residue analysis of ipronidazole and its metabolite at the 2 ppb level in turkey tissue, J.Agr. Food Chem. 19:1222 (1971).
19. J. Fellig, M. Kaykaty, L. Gonzales, H.G. Eisenbeis and R.E. Messersmith, Elimination of ipronidazole from the tissues of swine, Vet. Med. 70:31 (1975).
20. G. Weiss, N. Rose and P. Duke, Submitted to Xenobiotica.

21. P.N. Giraldi, G.P. Tosolini, E. Dradi, G. Nannini, R. Longo, G. Meinardi, G. Monti and I. De Carneri, Isolation identification and quantitative determination in humans of the metabolites of a new trichomonicidal agent (nimorazole), Biochem. Pharmacol. 20:339 (1971).
22. I.R. Flockhart, P. Large, D. Troup, S.L. Malcolm and T.R.Marten, Pharmacokinetic and metabolic studies of the hypoxic cell radiosensitizer misonidazole, Xenobiotica 8:97 (1978).
23. D.E. Schwartz, J-C. Jordan, W. Vetter and G. Oesterhelt, Unpublished results.
24. L.F. Zerilli, N. Rimorini, M. Landi, B. Cavalleri and A. Assandri, GC and MS in the identification of metabolites of 5-(1-methylethyl)-1-methyl-2-nitro-1H-imidazole in dogs, in: "Recent Developments in Mass Spectroscopy in Biochemistry and Medicine", A. Frigerio, ed., Plenum Publishing Corp., New York (1978).
25. A. Assandri, A. Perazzi, L.F. Zerilli, P. Ferrari and E. Martinelli, Metabolism of 5-isopropyl-1-methyl-2-nitro-1H-imidazole. Identification of some urinary metabolites in the dog, Drug Metab. Dispos. 6:109 (1978).
26. B. Cavalleri, G. Volpe, V. Arioli and G. Lancini, Synthesis and biological activity of two metabolites of 1-methy-5-(1-methylethyl)-2-nitro-1H-imidazole, an antiprotozoal agent, J. Med. Chem. 20:1522 (1977).
27. B. Testa and P. Jenner, Drug metabolism: Reduction of the nitrogen-containing functional groups, in: "Drugs and the Pharmaceutical Sciences" J. Swarbrick, ed., Vol.4, Marcel Dekker, New York and Basle (1976).
28. M. Mitchard, Bioreduction of organic nitrogen, Xenobiotica 1:469 (1971).
29. J.R. Fouts and B.B. Brodie, The enzymatic reduction of chloramphenicol, p-nitrobenzoic acid and other aromatic nitro compounds in mammals, J.Pharmacol.Exp.Ther. 119:197 (1957).
30. I. Bartosek, E. Mussini, C. Saronio and S. Garattini, Studies on nitrazepam reduction in vitro, Europ.J. Pharmacol. 11:249 (1970).
31. J.R. Gillette, J.J. Kamm and H.A. Sasame, Mechanism of p-nitrobenzoate reduction in liver: The possible role of cytochrome P-450 in liver microsomes. Mol.Pharmacol. 4:541 (1968).
32. R.T. Williams, "Detoxification Mechanisms". Chapman and Hall London (1959).
33. J. Reider and G. Wendt, Pharmacokinetics and metabolism of the hypnotic nitrazepam, in: "The Benzodiazepines", S. Garattini, E. Mussini and L.O. Randall, eds., Raven Press, New York (1973).
34. E. Eschenhof, Untersuchungen über das Schicksal des Antikonvulsivums Clonazepam im Organismus der Ratte, des Hundes und des Menschen. Arzneim.-Forsch. (Drug Res) 23:390 (1973).

35. K.M. Baker, M. Coerezza, L. Del Corona, A. Frigerio, G.G. Massaroli and G. Sekules, Metabolism of 1-methyl-5-nitro-2-(2'-pyrimidyl)imidazole, J.Pharmac.Sci. 63:293 (1974).
36. R.G. Fargher, Orientation of the nitro- and arylazo-glyoxalines. Fission of the glyoxalone nucleus, J.Chem. Soc. 117:668 (1920).
37. F.L. Pyman, Orientation of the 1-4 and 1-5 dimethylglyoxalines. Mode of fission of the 5-aminoglyoxalines, J. Chem. Soc. 121:2616 (1922).
38. R. Weidenhagen and R. Herrmann, Ueber das 4(5)-Amino-5(4)-methyl-imidazol, Berichte 68:2205 (1935).
39. M.A. Schwartz and M.K. Taylor (personal communication).
40. Y.C. Taylor and A.M. Rauth, Differences in the toxicity and metabolism of the 2-nitroimidazole misonidazole (Ro 7-0582) in HeLa and Chinese hamster ovary cells, Cancer Res. 38:2745 (1978).
41. P. Mazel, Determination of microsomal azo and nitro-reductase activities. in: "Fundamentals of Drug Metabolism and Drug Disposition", B.N. La Du, H.G. Mandel and E.L. May, eds., Williams and Wilkins, Baltimore (1972).
42. P. Populaire, F. Benazet, S. Pascal, G. Lebreton, B. Decouvelaere and L. Guillaume, Circulation et sort du métronidazole dans le tractus digestif chez le rat après administration par voie orale. Absorption due métronidazole par les muqueuses digestives. Thérapie 26:581 (1971).
43. R.R. Scheline, Metabolism of foreign compounds by gastro-intestinal microorganisms, Pharmacol. Reviews 25:451 (1973).
44. A.J.F. Searle and R.L. Willson, Metronidazole (FLAGYL): Degradation by the intestinal flora, Xenobiotica 6:457 (1976).

MECHANISM OF CYTOTOXICITY OF NITROIMIDAZOLES

D. I. Edwards, R. J. Knox, D. A. Rowley,
I. M. Skolimowski and R. C. Knight

Chemotherapy Research Unit, North East London Polytechnic
Romford Road, London E15 4LZ, England

INTRODUCTION

The nitroimidazole group of drugs are extensively used for the chemotherapy of anaerobic bacterial and protozoal disease and also for the radiosensitization of hypoxic tumours. In the treatment of hypoxic tumours, of particular clinical value are metronidazole (1-β-hydroxy-ethyl-2-methyl-5-nitroimidazole) and misonidazole (2-nitro-1-imidazolyl-3-methoxy-2-propanol). Both these drugs are important in that they exert a cytotoxic effect in addition to the radiosensitizing effects.

The primary target of the cytotoxic effect is DNA, both in hypoxic tumour cells and in anaerobic bacteria and protozoa and is a consequence of the reduction of the nitrogroup (2-5). In anaerobic pathogens reduction of the nitrogroup is mediated by redox processes of relatively negative potential including those involving ferredoxins and other iron-sulphur electron transport carriers, or by the pyruvate dehydrogenase complex (2,4,5,6). In hypoxic tumour cells the components responsible for nitro group reduction are unknown although it has been established that radiation causes nitro group reduction anoxically (7). Ferredoxin-linked reactions capable of reducing nitroimidazoles are inhibited vis-a-vis electron transport, but recover once all the drug is reduced (3). The biochemical lesion responsible for cytotoxicity is thus due to a diffusable reduction component and not transiently inhibited electron transport mechanism. In 1974 Ings et al. (8) suggested DNA as the target of cytotoxicity on the basis that metronidazole inhibits incorporation of labelled thymidine into the DNA of anaerobic protozoa and bacteria. Subsequent studies have shown that existing DNA is degraded in susceptible cells indicating this to be a general property of nitroimidazoles (9).

NITROIMIDAZOLE-INDUCED DNA DAMAGE

Initial attempts to characterize the effects of reduced nitroimidazoles on DNA involved in vitro reduction of the drug with dithionite under anoxic conditions and subsequent analysis of DNA damage. Such studies (10) showed that the midpoint of the helix-coil transition (the Tm value) was decreased and the melting range of DNA increased. In contract to the changes in the heating profile the cooling profile of DNA was unaffected. These results indicated that nitroimidazoles (metronidazole) did not act as intercalators, since these tend to increase the Tm value, nor did they act by cross-linking the DNA helix as do alkylating agents because the cooling curves were unaffected. The general effect of such drugs, which was confirmed by sucrose gradient sedimentation, was to cause strand breakage of DNA. However, in vitro experiments utilising dithionite are difficult to assess unambiguously since dithionite itself causes strand breakage (11).

In order to study the effect of reduced nitroimidazoles on DNA in the absence of chemical reducing agents our laboratory has developed an electrolytic reduction technique whereby the potential chosen for reduction is related to the polarographic halfwave potential ($E_{\frac{1}{2}}$) and corresponds to the plateau region of the wave of each drug. Provided the voltage is not more negative than -1.1V, which causes degradative conformational changes in DNA at the Hg electrode (12,13), the reduction may be carried out in the presence of DNA. Generally, the conditions chosen are those in which the drug-nucleotide ratio is 1:1 and corresponds to 20 μ moles of drug and 10 mg DNA in 67 ml 0.1 SSC buffer pH 7.0. Such conditions enable reduction to occur slowly during which period drug reduction can be monitored spectrophotometrically as a decrease in the extinction of the nitro group, and samples removed to assess the extent of DNA damage. No DNA damage is observed in oxic conditions, nor by the conditions of electrolytic reduction in the absence of drug. Consequently the effects described below may be solely attributed to the formation of reduced products of nitroimidazoles.

Damage to DNA may be observed by measuring a decrease in viscosity, Tm value, or in the ability of the molecule to renature after heating to the Tm value (14-17). Both viscosity and Tm determinations yield a general index of damage and reflect the integrity and stability of the helix. These effects may be attributed either to the possibility of DNA helix 'bending' (18) or to scission of the phosphodiester bonds, i.e. strand breakage. That the latter occurs may be assessed by hydroxyapatite chromatography, which measures the single strand content of DNA, by sucrose gradient sedimentation centrifugation which can detect single and double strand breaks, or by agarose gel electrophoresis which detects changes in the molecular weight of DNA. These techniques have revealed that the major effect of nitroimidazoles is to destabilize the DNA helix, lower the molecular weight average and increase the molecular weight range of DNA - all of which is caused by the formation of single and double strand breaks. Such effects are observed with misonidazole, metronidazole, 4,(5)-nitroimidazole, tinidazole, ornidazole, M&B 4998 (1-β-hydroxyethyl-4-nitro-pyrazole) and 8609RP (1,2-dimethyl-4-nitroimidazole)(15-17,19).

PROTECTION OF DNA DAMAGE BY AMINOTHIOLS

Aminothiols such as cysteamine are known to inhibit both the radiosensitizing and cytotoxic effects of nitroimidazoles (20). This effect also occurs in the in vitro electrolytic reduction system and cysteamine will not only protect against reduced drug-induced DNA damage, but will delay drug reduction. The degree of protection measured by those techniques mentioned above varies with the nature of the nitroimidazole. Thus cysteamine delays the reduction of metronidazole and 8609RP and protects DNA efficiently against DNA damage, but fails to delay reduction or protect against the effect of benznidazole. Figure 1 shows that the amount of protection irrespective of which techniques are used to measure DNA damage depends upon the difference in redox potential ($E_{1/2}$) of the drug and cystamine-cysteamine couple.

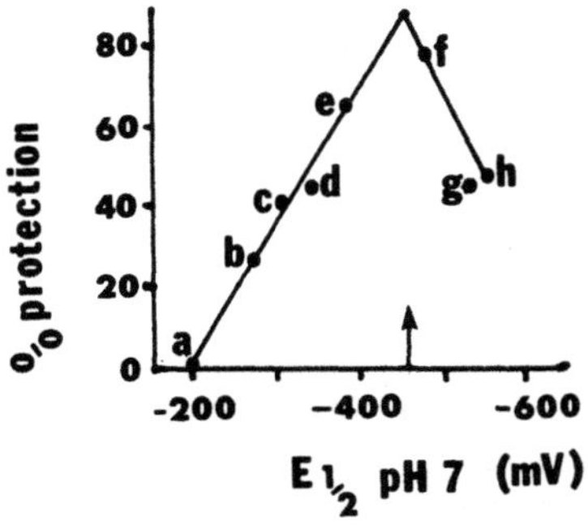

a = Benznidazole

b = Misonidazole

c = Dinitroimidazole (1st NO_2 group)

d = Ornidazole

e = Metronidazole

f = 8609 RP

g = Dinitroimidazole (2nd NO_2 group)

h = 4,(5)-nitroimidazole

Fig. 1. The protective effect of cysteamine against nitroimidazole-induced DNA damage measured as the decrease in relative viscosity of E.coli DNA at a drug-nucleotide ratio of 0.75 and a cysteamine-nucleotide ratio of 2. The arrow shows the $E_{1/2}$ of the cystamine-cysteamine couple.

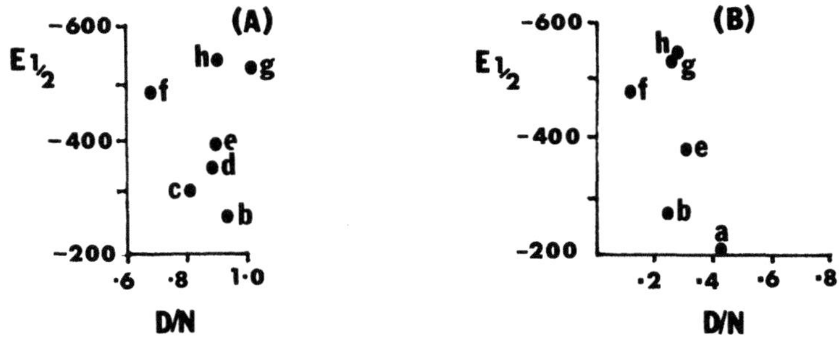

Fig. 2. Relationships between the reduction potential E½ in mV at pH 7 and DNA damage by reduced nitroimidazoles. (A) reduced nitroimidazole concentration (expressed as the reduced drug-nucleotide ratio D/N) to produce a 50% decrease in the relative viscosity of E. coli DNA. (B) reduced nitroimidazole concentration (expressed as the reduced drug nucleotide ratio) to produce a 50% increase in the single strandedness of E. coli DNA.

Thus, only those drugs having redox potentials close to that of cystamine will be efficiently protected. Whatever the mechanism of the protective effect it may be germane to the effectiveness of nitroimidazoles as chemotherapeutic agents.

MECHANISM OF DNA DAMAGE

The action of reduced nitroimidazoles in causing DNA damage depends upon the nature of the drug and the base composition of the DNA. Misonidazole, for example, damages DNA from Clostridium perfringens, calf thymus, Escherichia coli and Micrococcus lysodeikticus in decreasing order of efficiency, irrespective of which methods are used to assess DNA damage. These organisms have DNAs of % A + T 72, 60, 50 and 28 respectively, suggesting that the drugs do not cause strand breakage randomly, but that there is a specific target of the cytotoxic effect (21). The A + T content of DNAs from various sources may be directly correlated with DNA damage indicating that the target is A or T or both (21).

Recently the target has been identified as the thymine ring system of DNA (22). At reduced drug-nucleotide ratios of one, sufficient strand breakage and degradation of DNA occurs such that the products may be dialysed from DNA, separated and identified. Using Sephadex chromatography Knox et al. (22) have shown that the only products of DNA degradation released are thymine derived. At these drug concentrations the thymine is released (irrespective of the A + T content of the DNA) as d-thymidine-3-phosphate (32%), d-thymine-5-phosphate (22%), d-thymine-3,5-diphosphate (15%), d-thymine (18%) and thymine (3%). These values may vary by 5-10%. The amount of total thymine-derived products released is about 5% at drug-nucleotide ratio of one and correlates not only with DNA base composition but also the extent of DNA damage. This effect occurs both with misonidazole and metronidazole, and relative data are shown in Table 1. These nitroimidazoles do not damage poly (dG-dC) or apyrimidinic acid indicating that the effect is thymine specific. However, release of uracil-derived products is observed from RNA (R.J.K. and D.I.E., unpublished) suggesting that the specificity of the cytotoxic effect does not involve the 5-methyl group of thymine but the uracil nucleus. In this respect the specific nuclease action of nitroimidazoles resembles that of bleomycin and may well be of value in designing logical multi-drug combination therapy for hypoxic tumours.

The figure of about 5% thymine derived products released from DNA (with the exception of poly (dA-dT)) represents no statistically possible base sequence. Such results may be caused by a reduced species of relatively short half-life in which the time taken to find thymidine attached phosphodiester bonds is rate limiting. This may explain the constant release at relatively low A + T levels but would not hold for high levels, e.g. with poly (dA-dT). Such an explanation is purely conjectural at this time.

NATURE OF REDUCED NITROIMIDAZOLES

In 1978 Whitmore et al. (23) reported the presence of at least three reduction products of misonidazole obtained by chemical reduction and by exposure to hypoxic cells. These products were conjectured to be the fully reduced amine and intermediate reduction products. Electrolytic reduction of gram quantities of nitroimidazoles reveal considerably more than three reduction products as adduced by paper chromatography. Misonidazole produces seven reduced products which make up 90% of reduced material. Many of these are unstable in air and are highly deliquescent. Of these none correspond to the amine based on chromatographic, fluorescent and chemical criteria and we are unable to prepare a picrate derivative which would be possible if the amine was present. We have, however, tentative evidence for the formation of methyl

Table 1. Reduced Misonidazole-induced Release of Thymine-derived Products from DNA

DNA	% A & T	Relative (a) damage	% thymine (d) in dialysate	% total thymine released from DNA (b)
Micrococcus	28	0.40	84	5.1
E. coli	50	0.71	86	5.0
E. coli[(c)]	50	-	-	4.9
Calf thymus	60	0.86	83	5.1
Clostridium	72	1.13	85	5.6
poly (dA-dT)	100	7.10	95	25.0

(a) relative damage is expressed as the relative integrated peak areas of eluted dialysable material, see ref. 22; (b) includes all thymine derived products as mentioned above; (c) data for single stranded DNA; (d) includes all thymine-derived products.

acetamide and dimethyl acetamide from 8609 RP and dimetridazole (1,2-dimethyl-5-nitroimidazole) respectively based on infra red spectra, and the presence of a characteristic odour. Acetamide has also been detected as a product of metronidazole reduction by intestinal microflora (24) and suggests that the imidazole ring undergoes fission.

A different approach to the study of drug reduction is the utilization of a coulombmeter in the electrolytic reduction circuit. This enables the measurement of the apparent number of electrons (n app) involved in the reduction process to be measured. As an example misonidazole is completely reduced with an electron requirement of 4.01 indicating that a major reduction product would be the hydroxylamine. This has also been detected by HPLC post-column derivatization techniques (P. Wardman, personal communication).

Table 2 indicates that misonidazole is reduced to the hydroxylamine and suggests that the other six reduction products arise from decomposition of this product. Metronidazole gives a non-integral value which may represent a distribution of 2, 4 and 6-electron reduction products since 4(5)-nitroimidazole readily yields the amine (6-electrons). In contrast, in the presence of DNA at a drug nucleotide ratio of one the n app. value changes. It should be noted that in these experiments a change in coulombmeter reading of ca. 58,000 is required for a 1-electron addition and consequently the non-integral values are significant. With the exception of metronidazole it appears that, because n app. decreases in the

Table 2. Electron Requirements for Nitroimidazole Reduction

Compound	$E_{1/2}$,V(a)	Reduction potential V	n app.	Comments
Misonidazole	-0.272	-0.8	4.01	pH 6.8-8
" and DNA	-0.272	-0.8	3.88	pH 6.6-10
Metronidazole	-0.382	-0.9	3.51	pH 7 - 10
" and DNA	-0.382	-0.9	3.68	
4 (5) nitroimidazole	-0.545	-1.0	6.08	pH 7.1-8.4
" and DNA	-0.545	-1.0	5.50	pH 6.9-10.5
2-nitroimidazole (Azomycin)	-0.375	-0.9	3.80	
2,5-dinitroimidazole and	-0.308 -0.535	-1.0	8.0	pH 6.0-8.5

(a) The polarographic half wave potentials ($E_{1/2}$) ane electrolytic reduction potentials are those based on an $Ag/AgCl_2$ reference electrode and differ from the standard calomel electrode by approximately -40 mV. $E_{1/2}$ values are quoted at pH 7.0.

presence of DNA, damage may be caused by electron donation from DNA to the active reduction intermediate. The reduced intermediate may either subsequently decompose or undergo further reduction as shown below

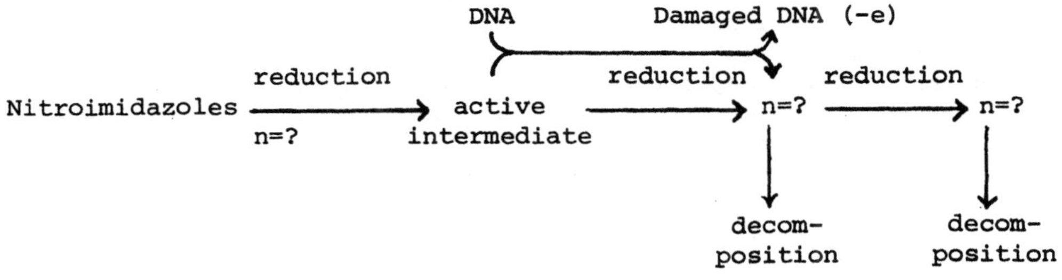

The anomalous behaviour of metronidazole in having an increase in n app. in the presence of DNA may be explained if the active intermediate rapidly decomposed to give products unable to undergo further reduction.

The nature of the active reduction intermediate is unknown but it is unlikely to be a 4- or 6-electron product since the final stable reduction products show no cytotoxicity towards Clostridium or E. coli when grown anaerobically, or E. coli aerobically. A likely candidate may be the 1-electron radical anion, the 3-electron nitroso radical, which requires proton addition to enable further reduction to occur, and would therefore be relatively stable or the nitroso derivative itself. The imidazole radical which is formed together with the nitrite ion both by γ-radiolysis and electrolytic reduction (A. Breccia, personal communication) is an unlikely candidate since its formation would not alter n app. in the presence of DNA.

CONCLUDING REMARKS

It has already been established that a correlation exists between the 1-electron redox potential of nitroimidazoles and their radiosensitizing efficiency as well as their aerobic cytotoxicity (25,26). Following the demonstration that misonidazole shows a greater cytotoxic effect in hypoxic than oxic cells (27)attempts have been made to correlate hypoxic cytotoxicity with the 1-electron redox potentials. Cytotoxicity of nitroimidazoles is dependent upon different parameters from radiosensitization. Thus, in cells the effect is slow compared to radiosensitization and appears to be temperature-dependent (28). The differences may lead one to expect that hypoxic (or anaerobic) cytotoxicity may not be correlated with 1-electron redox potentials but may be related if the formation of the 1-e radical anion is rate limiting. Figure 2 shows that there appears to be no apparent correlation between the drug concentration (expressed as the drug-nucleotide ratio) and the $E^{1/2}$ of a variety of nitroimidazoles (which is closely related to the 1-e redox potential) for two major biological effects of the cytotoxic mechanism viz: the drug concentration required to produce a 50% decrease in the relative viscosity of DNA, and the concentration required to produce a 50% increase in the single strandedness of DNA.

It should be emphasized that in correlations of the anaerobic or hypoxic cytotoxic effect the concentration of reduced drug must be taken into account. Otherwise, if relative rates of drug reduction are ignored an automatic correlation with 1-electron redox potentials will be found since this governs the ease or difficulty of drug reduction. The results shown in Fig. 2 indicate that the cytotoxicity of nitroimidazoles at the biological target is independent of the redox potential and is of approximately equal activity once the active reduced species has been formed, irrespective of the nature of the drug molecule from which it derived. Further studies to substantiate or refute this possible mechanism is necessary to establish which occurs in hypoxic cells.

ACKNOWLEDGEMENTS

We gratefully acknowledge grants from the Cancer Research Campaign, Medical Research Council and the Science Research Council. R.C.K. is a Research Fellow of the Cancer Research Campaign and D.A.R. is a Science Research Council CASE postgraduate student. We thank May and Baker Ltd for supplying metronidazole, dimetridazole, 8609 RP, and M&B 4998, Roche Products Ltd for misonidazole and Professor K.C. Agrawal for 2,5-dinitroimidazole.

REFERENCES

1. G. E. Adams, in: "Treatment of Radioresistant Cancers", M. Abe, K. Sakamoto and T. L. Phillips eds., Elsevier/ North-Holland, Amsterdam (1979) pp.3-19.
2. D. I. Edwards, M. Dye, and H. Carne, J. Gen. Microbiol. 63: 297 (1973).
3. R. W. O'Brien, and J. G. Morris, Arch. Mikrobiol. 84:225 (1974).
4. D. I. Edwards, and A. I. Schoolar, Zeit. Pflanzenphysiol. 64:73 (1971).
5. D. I. Edwards, G. E. Mathison, and D. J. Platt, Zeit. Pflanzenphysiol. 71:424 (1974).
6. M. L. Britz, and R. G. Wilkinson, Ant. Ag. Chemother. 16:19 (1979).
7. R. L. Willson, W. A. Cramp, and R. M. J. Ings, Int. J. Radiat. Biol. 26:557 (1974).
8. R. M. J. Ings, J. A. McFadzean, and W. E. Ormerod, Biochem. Pharmacol. 15:1421 (1974).
9. C. W. Plant, and D. I. Edwards, J. Antimicrob. Chemother. 2:203 (1976).
10. D. I. Edwards, J. Antimicrob. Chemother. 3:43 (1977).
11. R. Cone, S. K. Hasan, J. W. Lown, and A. R. Morgan, Canad. J. Biochem. 54:219 (1976).
12. V. Brabek, and E. C. Palecek, Stud. Biophys. 60:105 (1976).
13. H. W. Nurnberg, and P. Valenta, Colston Papers 29:201 (1978).
14. D. I. Edwards, R. C. Knight, and I. Kantor, in: "Current Chemotherapy", W. Siegenthaler and R. Luthy eds., American Society for Microbiology, Washington DC (1978) pp.714-716.
15. D. A. Rowley, R. C. Knight, I. M. Skolimowski, and D. I. Edwards, Biochem. Pharmacol. 28:3009 (1979).
16. R. C. Knight, I. M. Skolimowski, and D. I. Edwards, Biochem. Pharmacol. 27:2089 (1978).
17. R. C. Knight, D. A. Rowley, I. M. Skolimowski, and D. I. Edwards, Int. J. Radiat. Biol. 36:367 (1979).
18. K. E. Reinert, J. Molec. Biol. 72:593 (1972).
19. D. I. Edwards, D. A. Rowley, R. J. Knox, I. M. Skolimowski and R. C. Knight, in: "Current Chemotherapy and Infectious Disease", American Society for Microbiology, Washington, USA (1980) (in press).

20. E. J. Hall, M. Aster, C. Geard, and J. Biaglow, Br. J. Cancer 35:809 (1977)
21. D. A. Rowley, R. C. Knight, I. M. Skolimowski, and D. I. Edwards, Biochem. Pharmacol. (1980) (in press).
22. R. J. Knox, R. C. Knight, and D. I. Edwards, IRCS J. Med. Sci. 8:190 (1980).
23. G. F. Whitmore, S. Gulyas, and A. J. Varghese, Br. J. Cancer 37:Suppl.III:115 (1978).
24. R. L. Koch, E. J. T. Chrystal, B. B. Beaulieu, and P. Goldman, Biochem. Pharmacol. 28:3611 (1980).
25. P. Wardman, in: "Radiosensitizers of hypoxic cells", A. Breccia, C. Rimondi, and G. E. Adams eds., Elsevier/North-Holland, Amsterdam (1979) pp. 91-108.
26. G. E. Adams, E. D. Clarke, I. R. Flockhart, R. S. Jacobs, D. S. Sehmi, I. J. Stratford, P. Wardman, M. E. Watts, J. Parrick, R. G. Wallace, and C. E. Smithen, Int. J. Radiat. Biol. 35:133 (1979).
27. R. M. Sutherland, Cancer Res. 34:3501 (1974).
28. I. J. Stratford, in: "Radiosensitizers of Hypoxic Cells", A. Breccia, C. Rimondi, and G. E. Adams eds., Elsevier/North-Holland, Amsterdam (1979), 109.

NITROIMIDAZOLES AS ANTIPARASITIC AGENTS

I. de Carneri

Department of Parasitology
University of Pavia
Italy

The nitroimidazoles were launched on their pharmacological career in 1956 when azomycin or 2-nitroimidazole, a natural product obtained by fermentation, was found active against certain bacteria and *Trichomonas*. For a decade, nitroimidazole derivatives were used only to combat protozoa of the Trichomonadida order(1)(including *Histomonas meleagridis* in veterinary medicine), and for Vincent's angina, a spirochaetal infection. Subsequently their use was extended to other parasitic infections (2), mainly amoebiasis and giardiasis, and only in the last five years have they been definitely employed in bacteriology (3)(Fig. 1) and as radiosensitizers.

In the field of antiparasitic therapy, I should like to distinguish two periods in the history of the nitroimidazoles.

1) The first period started at the end of the fifties, and had reached its climax in the sixties when for the first time drugs became available for both vaginal and extra-vaginal localizations of *Trichomonas vaginalis*, such as the urethra and Bartholin's and Skene's glands, not reached by topical treatments. Such drugs are easily absorbed after oral administration, guarantee high levels of active substances in vaginal fluids (Fig. 2), are excreted mainly in microbiologically active form by the kidneys and pass through the urethra. For the time being, these drugs are irreplaceable, because intensive research for alternative chemical structures has failed to produce anything worthwhile. Research for new trichomonacidal nitroimidazoles (an economically profitable field) has been continued, although less vigorously, throughout the seventies, but no decisive further improvements have been achieved. In fact, we do not consider an advantage the prolongation of the half-life obtained with certain new 5-nitroimidazoles with respect to, say, nimorazole,

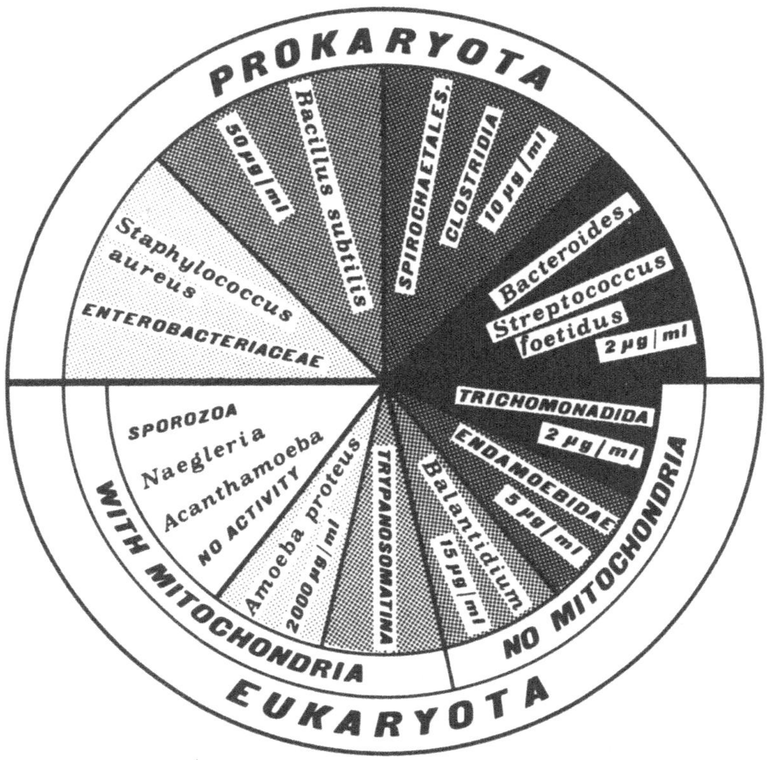

Fig. 1. Activity spectrum of nitroimidazoles. (The activity of nimorazole is given in µg/ml.)

which despite its very short half-life guarantees good clinical results after a single dose. In our mind, nitroimidazoles for such mild parasitoses as trichomoniasis and giardiasis should act promptly (Figs 3-8) and be excreted rapidly in order to limit side-effects. These are features suitable for "hit and run" treatment tactics (4). There are fortunately no immediate problems of "wear" on these drugs due to resistance arising in clinical practice because, though resistance of Trichomonas vaginalis can easily be induced in vitro and in experimental infections, as we reported several years ago (5,6), it appears to be still rare in human infection (7). Besides, it is now confirmed that other nitroimidazoles, with different physicochemical properties, preserve various degrees of activity against resistant strains (Figs 9,10,11).

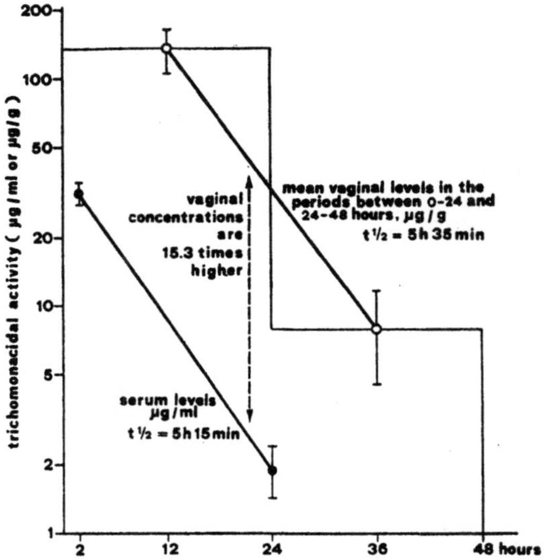

Fig. 2. Oral administration of 2g nimorazole to 4 healthy female subjects. Average levels of activity against <u>Trichomonas vaginalis</u> in serum and vaginal secretion, collected by Tampax plugs for 24 hours, extracted with methanol and determined <u>in vitro</u> (3).

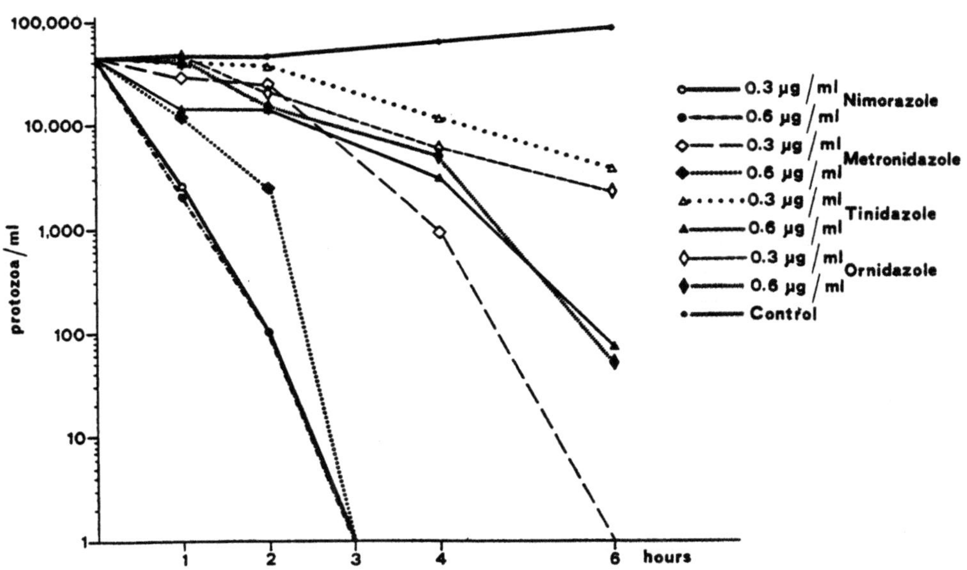

Fig. 3. Count of colonies of <u>Trichomonas vaginalis</u> in CPLM agar after contact with some nitroimidazoles in fluid CPLM medium at 37°C.

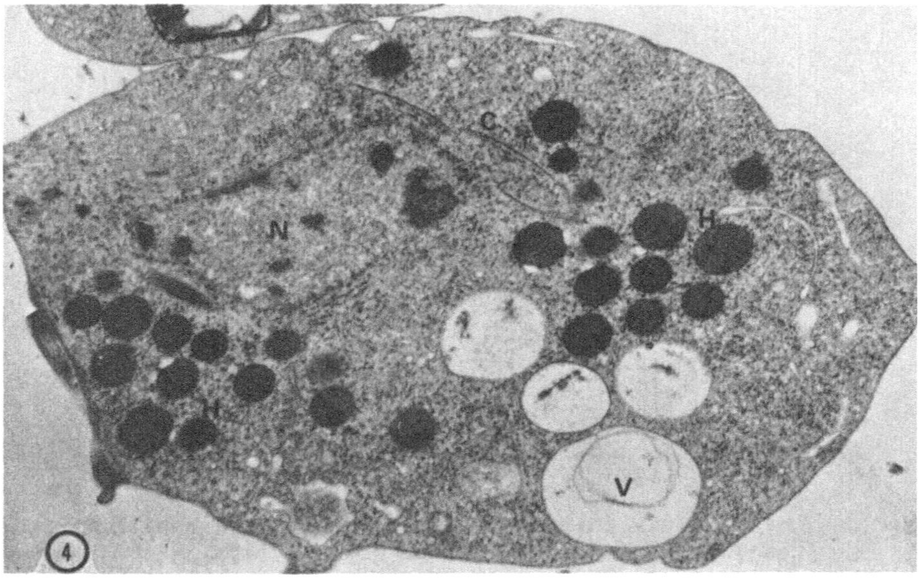

Fig. 4. *Trichomonas vaginalis* in culture; a number of electron-dense hydrogenosomes (H), vacuoles (V), part of the costa (C) and the nucleus (N) are visible. X 9,000.

The use of the same drugs in giardiasis is based on their marked specific activity (we have found nimorazole active *in vitro* on pure cultures of Giardia intestinalis, Fig. 12, at a concentration of 0.5 µg/ml for 24 h, Fig. 13 (8,9) and on the fact that almost all the alternative drugs present toxicity problems. After the treatment *in vitro* of *Giardia* trophozoites with nitroimidazoles, the most conspicuous observable phenomena are the appearance of pseudomyelinic figures in a swollen cytoplasm (Figs 14,15,16), which subsequently empties (Fig. 17), and, on the other hand, alteration of the cell profile, which becomes undulated (Fig. 18). The projections forming on the cell surface probably peduncularize and detach, forming half-moon shaped bodies peripheral to the cell (Fig. 19) - which we have described elsewhere for *Giardia muris* (10). In man, required treatment ranges from a single dose to 2-3 days therapy. Longer treatments should be avoided, and in such cases paromomycin, for example, appears preferable.

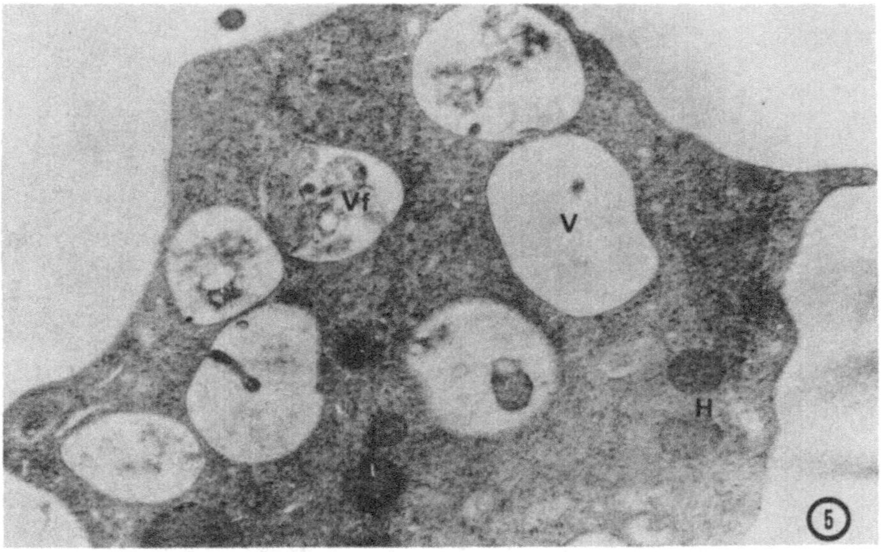

Fig. 5. Trichomonas vaginalis treated in vitro for 3 hours with 6 µg/ml of Nimorazole. The cytoplasm is pasty and abnormally dilated; phagocytosis vacuoles (Vf) are evident. V = vacuole, H = hydrogenosomes.

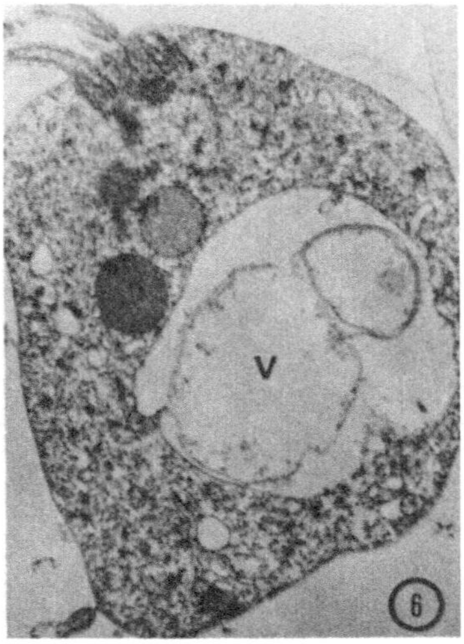

Fig. 6: Trichomonas vaginalis treated for 6 hours with 6 µg/ml of Nimorazole. The cytoplasmic matrix is empty. A swollen vacuole (V) is evident. x 7,800.

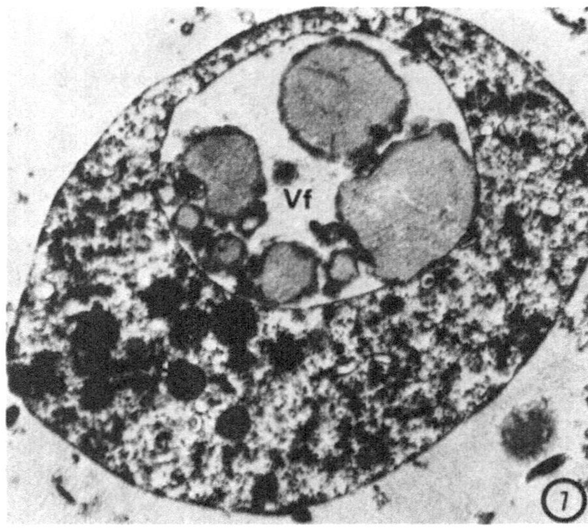

Fig. 7. Trichomonas vaginalis treated in vitro for 8 hours with 6 µg/ml of Nimorazole. The phagocytosis vacuoles have grown abnormally. Some hydrogenosomes (H) are almost intact. x 8,000.

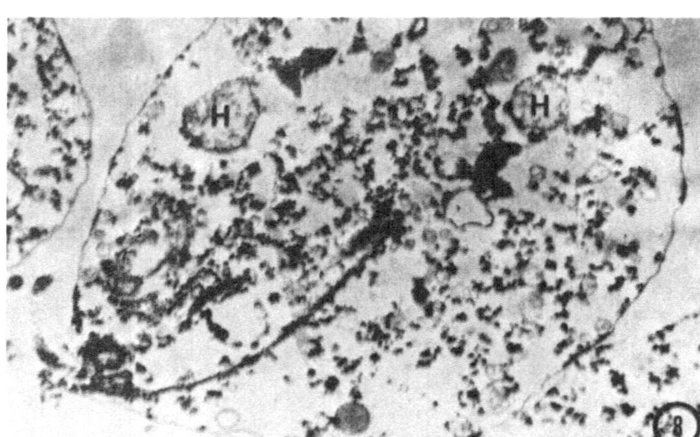

Fig. 8. Trichomonas vaginalis treated in vitro for 6 hours with 9 µg/ml of Nimorazole. The cell is completely damaged. We can only recognize some cytoplasmic residuals, as "ghosts" of hydrogenosomes (H). x 6,300.

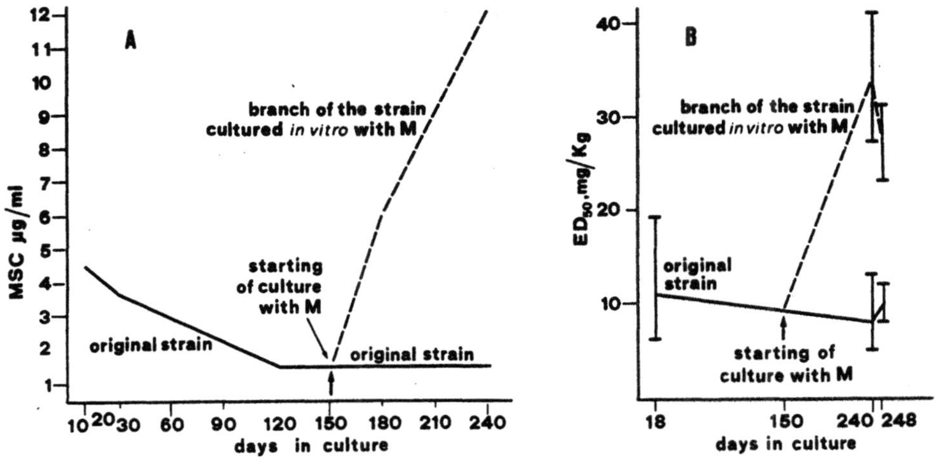

Fig. 9. A. Minimum sterilizing concentrations (MSC) in vitro of metronidazole on Trichomonas vaginalis after many passages in CPLM medium (——) and in the same medium with increasing concentrations of M (metronidazole - - -).

B. Effective Dose$_{50}$ (with 95% fiducial limits) in mice infected with the two branches of the strain of Trichomonas vaginalis described in Fig. A.

In amoebiasis, it is not wise to shorten treatment this way, and larger doses of metronidazole, nimorazole, tinidazole, etc. are required. The use of nitroimidazoles is thus only fully justified in invasive intestinal amoebiasis and in extraintestinal forms. For amoebic infections confined to the intestinal lumen, as most are, other drugs can be used, e.g. dichloroacetamide derivatives. In particular, long-term daily use of nitroimidazoles for months on end, as has been done in Mexico for chemoprophylaxis of amoebiasis (11) is not justified.

A dampening of the old enthusiasm for nitroimidazoles against amoebiasis is noticeable in the USA (12), where the FDA seems also to be cautious in allowing registration of new analogs for treatment of trichomoniasis.

The use of nitroimidazoles is fully justified in balantidiasis, a severe condition in which these drugs, nimorazole in particular (Fig. 20), have proved their worth in Mexico (13) and Colombia (14).

In cutaneous and mucocutaneous leishmaniasis, on the other hand, the hopes raised by reports some years ago on the efficacy of

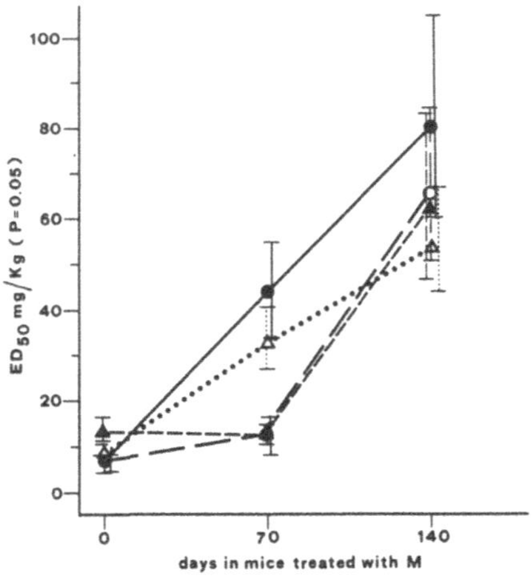

Fig. 10. Induction of resistance in 4 strains of Trichomonas vaginalis in 1971 (6) by means of subcutaneous passages in mice treated with increasing doses of Metronidazole. At the end these strains became resistant in similar tests *in vivo*, but not in normal tests *in vitro*, in anaerobiosis. This fact was explained only years later (7).

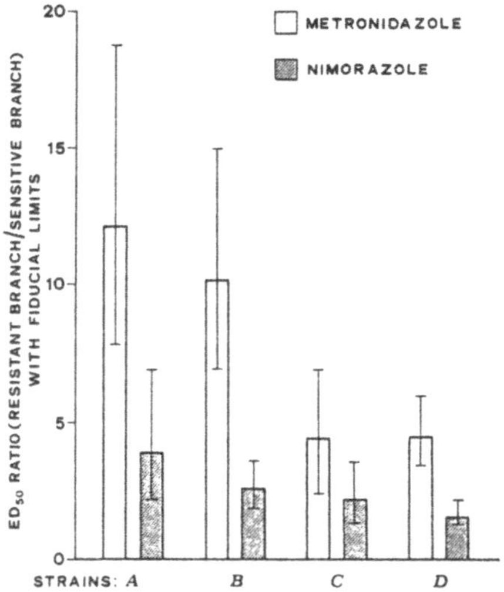

Fig. 11. ED_{50} variations of metronidazole and nimorazole on 4 strains of *T. vaginalis* made resistant to Metronidazole by subcutaneous transfers in mice. Plotted on the ordinates are values and confidence limits (for P = 0.05) of the ratios between activities on the resistant and on the sensitive branch in mice (ref. 6).

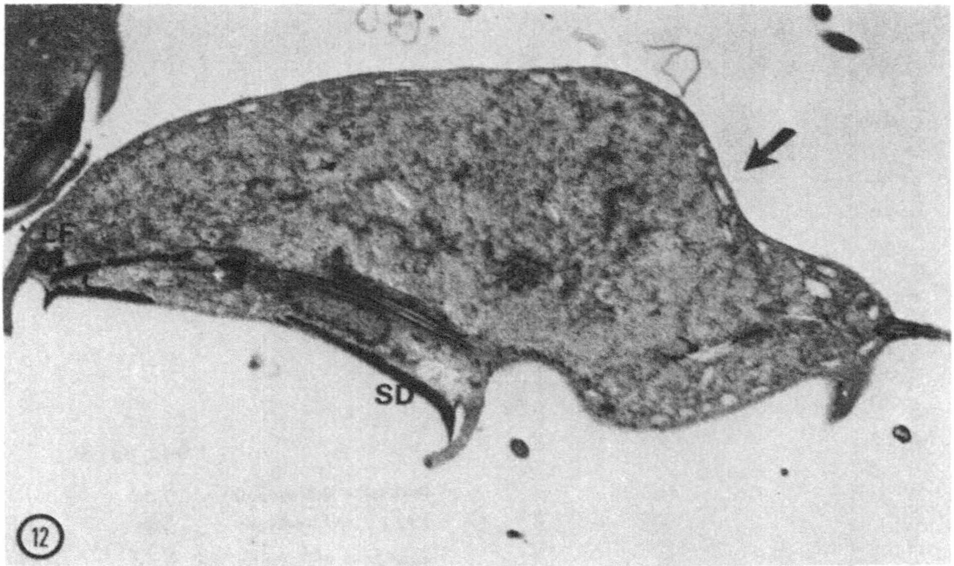

Fig. 12. Section of Giardia intestinalis cultured in Diamond's medium, illustrating elements of the "sucking disc" (SD), dorsal cisternae (↙) and a lateral flagellum (LF). x 8,500.

the above mentioned, basically anti-trichomonad drugs seem not to have been founded. Similarly, these classic nitroimidazoles have only marginal activity against nematodes. It is still debated whether metronidazole is really effective against Dracunculus medinensis, but whatever the verdict, after treatment with metronidazole the worm must still be extracted by traditional methods (15). But this, I believe, is where we move into the second phase of the story of nitroimidazoles as antiparasitic agents.

2) The second period is marked by research for nitroimidazoles active against other severe, but rare parasitoses which promise no economic rewards for the pharmaceutical firms engaged in this type of research. This field of research is concerned with American trypanosomiasis (Chagas' disease) and African trypanosomiases (sleeping sickness, trypanosomiases of cattle) as well as various forms of leishmaniasis and filariasis. At this point we may look back to the fact that some nitroimidazoles with a certain therapeutic promise against the trypanosomiases had already been identified in the sixties. Among the compounds that may be of interest in this respect are some 2-nitroimidazoles, a category that was difficult to synthesise until a few years ago. This, and the lack of interest in development of drugs against parasitoses prevalent among the less

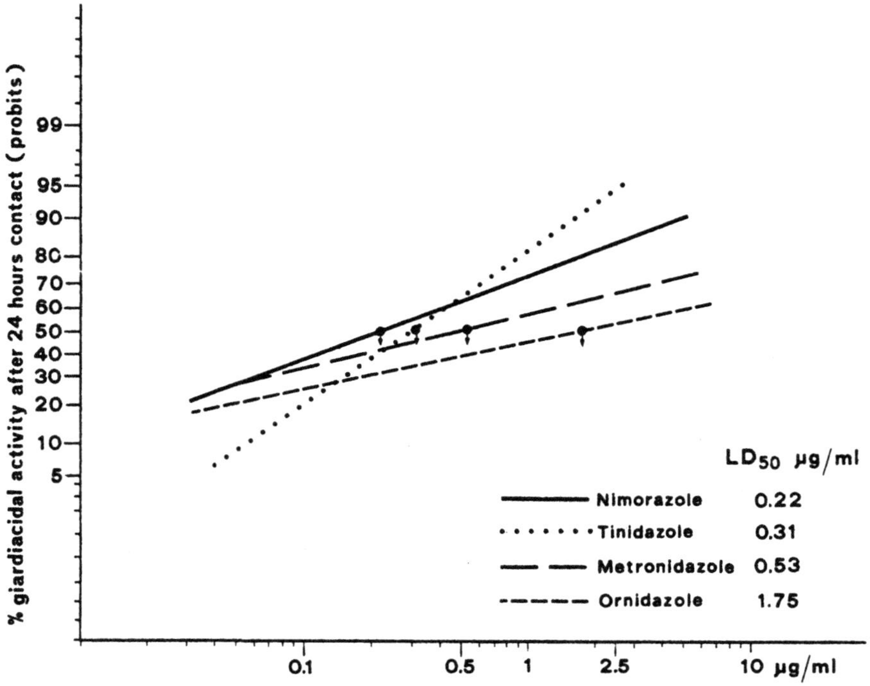

Fig. 13. Protozoocidal activity of some nitroimidazoles on pure cultures of Giardia intestinalis (5.000 protozoa/ml) after 24 hours contact at 37°C in Diamond's medium.

affluent peoples of the Third World, have hampered this line of research for more than ten years (with few exceptions: benznidazole, active on Trypanosoma cruzi, went through 10 years of laboratory studies, then field trial in South America).

The catalyzing effect of WHO, now backed by World Bank funds for the Special Programme for Research and Training in Tropical Diseases, has given new courage only recently to the few pharmaceutical manufacturers who did not drop research on such antiparasitic drugs. In connection with this Programme, centres like those of the Liverpool School of Tropical Medicine, then directed by Prof. W. Peters, and of the University of Georgia in Athens, USA, directed by Prof. W. Hanson, had the courage to pursue the task, not strictly academic, of screening new antiparasitic agents, in this case against leishmaniasis. As a result of these initiatives, we hope that the new nitroimidazoles that hold out some promise as effective trypanocidal or antifilarial agents will no longer be considered "orphan drugs" in the sense suggested by Lasagna (16,17).

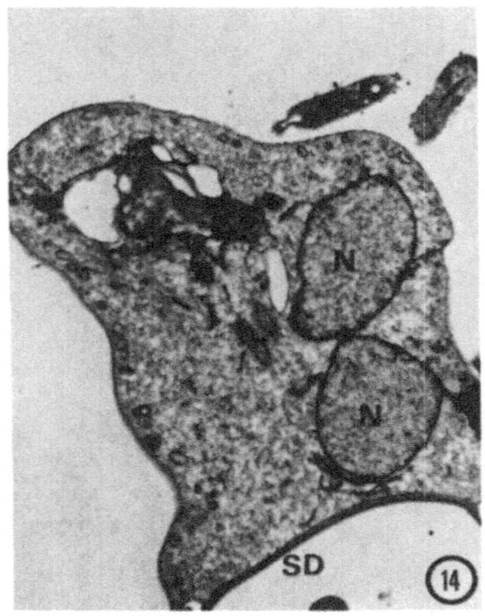

Fig. 14. Section of *Giardia intestinalis* treated *in vitro* for 1½ hours with 30 µg/ml of Nimorazole. Note the swelling of the cell. Nucleus (N), "sucking disc" (SD). x 9,000.

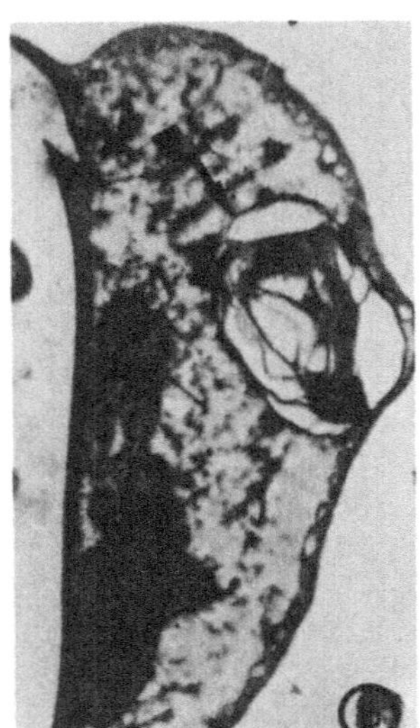

Fig. 15. Section of *Giardia intestinalis* treated *in vitro* for 1½ hours with 60 µg/ml of Nimorazole. In the partially emptied cytoplasmic matrix, cytolisosomes appear (↓). x 15,000.

Fig. 16. Enlargement of Fig. 15. Note the winding and the adhesion of the membranes (↖). x 28,000.

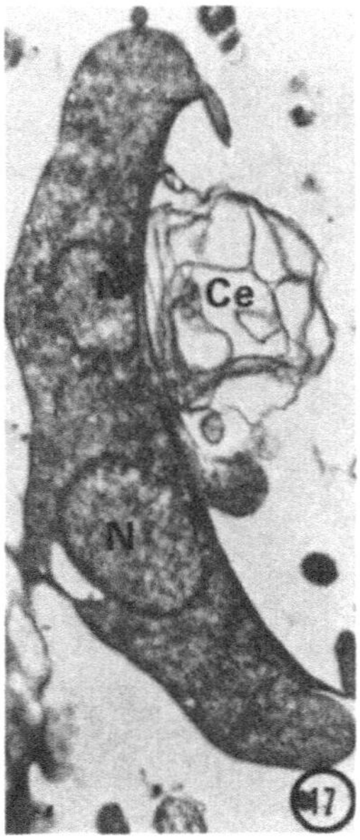

Fig. 17. Section of Giardia muris treated in vitro for 1½ hours with 60 μg/ml of Nimorazole. Cytoplasmic extrusions (Ce) are evident. x 12,000.

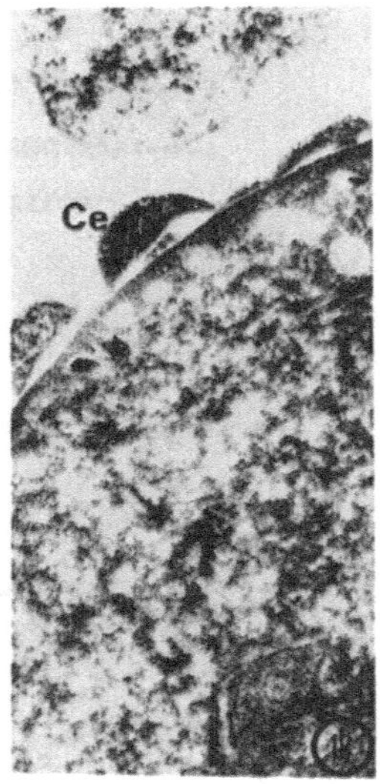

Fig. 18. Section of Giardia muris treated in vitro for 1½ hours with 90 μg/ml of Nimorazole. Note the cytoplasmic extrusions (Ce) half-moon shaped on the cell surface. x 27,000.

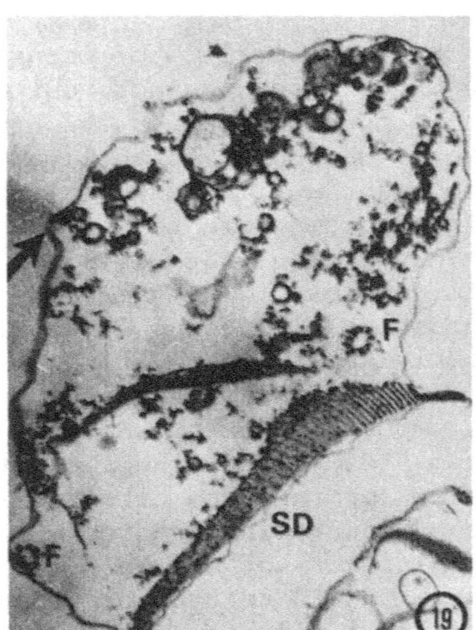

Fig. 19. Section of Giardia muris treated in vitro for 1½ hours with 90 μg/ml of Nimorazole. The cell is destroyed; the cytoplasmic matrix is deeply damaged. Only the "sucking disc"(SD) and flagella(F) are still visible. x 20,000.

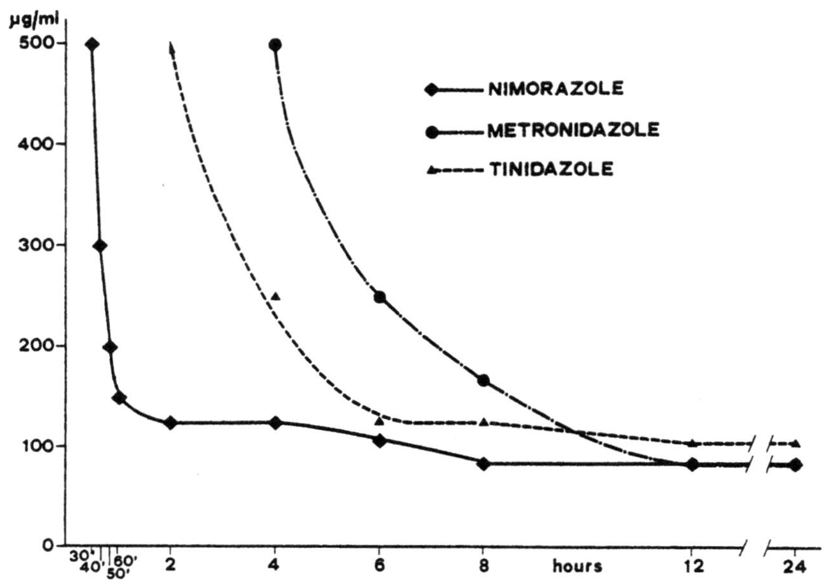

Fig. 20. Speed of protozoocidal activity of some nitroimidazoles on monophasic cultures of Balantidium coli (3.000 protozoa/ml) at 37°C.

Biochemical knowledge in this field of parasitology is scant, and as studies of these drugs make progress, our general knowledge increases too. This already happened for the hydrogenosomes of Trichomonadida, and for the anaerobic metabolism in trichomonads and Entamoeba, responsible for activation of nitroimidazoles in these microorganisms. The Trypanosomatidae to not possess ferredoxin. One of there enzymes, α-glycerophosphate oxidase, might reduce and activate certain nitroimidazoles. Our understanding of the life cycles of African trypanosomes also gains from this therapeutic research. A certain activity of a derivative of 5-imidazole-hexamethylene-isoxazoline (18) against Trypanosoma brucei in the central nervous system is perhaps due to the presence in the brain of amastigotes of this species (though their existence has not yet been demonstrated) able to activate the nitroimidazoles. Similarly, Malanga et al. (in press), continuing studies started more than 10 years ago (19), found an activity of some nitroimidazoles on the amastigotes of Trypanosoma cruzi, responsible for Chagas' disease. From the first laboratory tests in mice, also Fexinidazole (20), one of the many nitroimidazoles synthesized by Winkelmann (Arzneimittel-Forschung, between 1977 and 1978), seems to be of some interest. Benznidazole, a 2-nitroimidazole with a minimal trypanosomicidal concentration of 3-6 µg/ml in tissue cultures, is now used in South America for the treatment of this condition. However, de Maar (21)

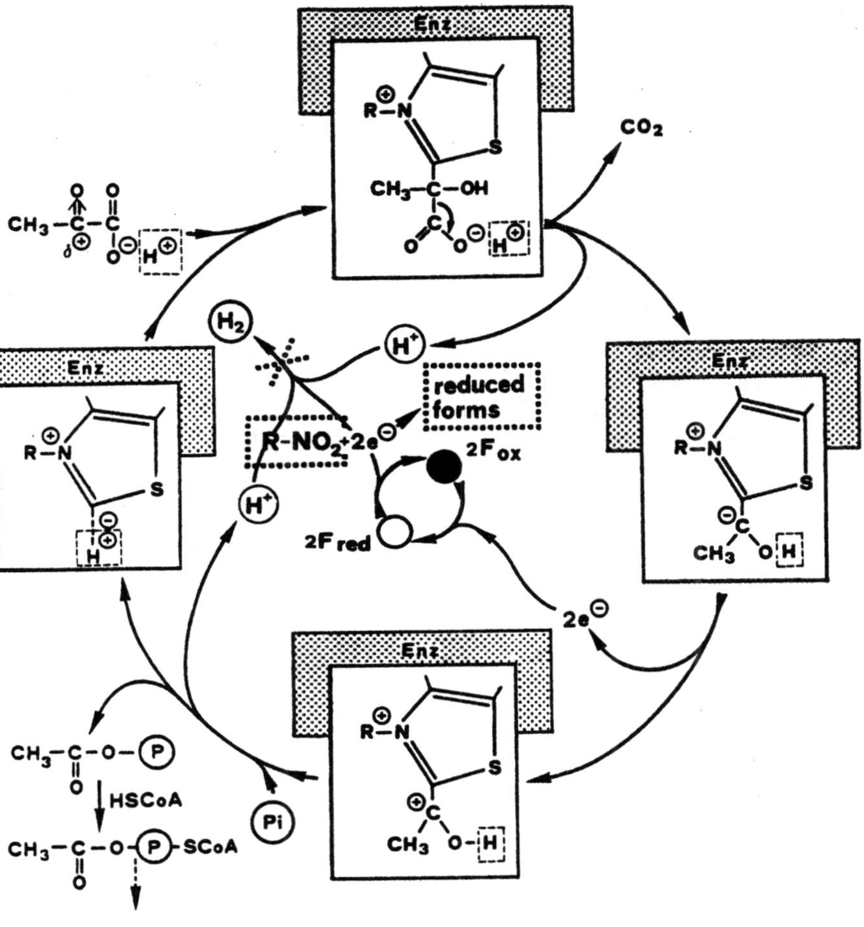

Fig. 21. Simplified scheme of hydrogen gas liberation in some anaerobic microorganisms and interference of nitroimidazoles, which are reduced to active compounds. F = ferredoxin or ferredoxin-like enzymes.

of WHO warns that this drug needs proof of sufficient activity against tissue forms in humans and has problems with patient acceptance (at daily doses of 5-7 mg/kg for one month). Objectively, therapy of chronic Chagas' disease is an extremely difficult task (see the empty space left by "The Medical Letter, 1979, for drugs against this condition)(22). Besides, for mass campaigns directed to the 12 millions of mainly rural South American patients, simple and short schemes of treatment are needed.

The 2-nitroimidazoles are now being studied under the auspices of the UNDP/World Bank/WHO's Special Programme for Research and

Training in Tropical Diseases, as potential filaricides. They are still at the stage of laboratory screening.

According to Polak et al.(23) the 2-nitroimidazoles act by means of a direct mechanism, different from that responsible for the action of the 5-nitroimidazoles, which is based on their intracellular reduction to cytotoxic compounds. Should this be so, then for these 2-nitroimidazoles there would be different, and promising, pharmacological and toxicological prospects. But the observations published by Goldstein et al.(24) indicate that the 2-nitroimidazoles' nitrogroup must also be partially reduced by target organisms.

In summary, the rules of the game for finding new nitroimidazoles for use in parasitology seem obvious enough - to identify substances with a redox potential such that the protozoon or metazoon in question can reduce them (Fig. 21), but which will not be reduced by normal cells of the human body. The candidate compounds must also be well absorbed and distribute appropriately after oral administration; its half-life and length of treatment should place it mid-way between the opposing requirements of hitting the parasite and not harming the host, and it should not present other toxicity or safety problems. The task is not easy. As is so often the case in the chemotherapy of parasitic infections, here again we are apparently caught between Scylla and Charybdis.

However, some new epidemiological evidence is now available which indicates that any risk of cancer in humans is not perceptible (25).

1. I. de Carneri, Antiprotozoan activity of nitroimidazoles, Arzneimittel-Forschung, 19:382.
2. I. de Carneri, A. Bianchi, S. Castellino, G. Meinardi, and F. Trane, New data on the mechanism and speed of action of nitroimidazoles on prokaryota and eukaryota with and without mitochondria, Arzneimittel-Forschung, 26:765 (1976).
3. I. de Carneri, Taxonomy of amoebae and antiprotozoal activity of nitroimidazoles and dichloracetamide derivatives. Proceedings of the International Conference on Amebiasis, Centre de Estudios sobre Amibiasis, Mexico City, October 27-29, 1975. p.174.
4. I. de Carneri, La trichomonase uro-génitale mérite-t-elle un traitement systémique? Bulletins et memoires de la société de médécine de Paris, 181:189 (1978).
5. I. de Carneri, Variation of the sensitivity of a strain of Trichomonas vaginalis to metronidazole after culturing in the presence or absence of the drug, in: "Proceedings of the First International Congress of Parasitology", A. Corradetti, ed.,

Pergamon Press-Tamburini Editore, Milan (1964) p.366.

6. I. de Carneri and F. Trane, In vivo resistance to metronidazole induced on four recently isolated strains of Trichomonas vaginalis, Arzneimittel-Forschung, 21:377 (1971).
7. M. Muller, Nitroimidazole resistance in anaerobic bacteria and protozoa, Conference on nitroimidazoles, Cesenatico, 27-30 August, Program and Abstracts, 35 (1980).
8. F. Trane and I. de Carneri, Attività del nimorazolo in vitro su culture pure di Giardia intestinalis, 10th Congresso della Società Italiana di Parassitologia, Alghero 20-23 May 1980, Parassitologia, in press.
9. M. de Eguileor, F. Trane and I. de Carneri, Aspetti ultrastrutturali dell'azione dei nitroimidazoli su culture pure di Giardia intestinalis, 10th Congresso della Societa Italiana di Parassitologia, Alghero 20-23 May 1980, Parassitologia, in press.
10. I. de Carneri, M. D. Candia Carnevali and F. Trane, Aspetti ultrastrutturali dell'azione del nimorazolo su Giardia muris, Giornale di Malattie Infettive e Parassitarie, 28:503 (1976).
11. R. Delgado y Garnica and E. Martinez-Murray, Quimioprofilaxis de la amibiasis con metronidazol. Rev. Inst. Med. Trop. São Paulo, 13:307 (1971).
12. W. Modell, "Drugs of choice 1980-1981". Mosby Co., St. Louis (1980).
13. F. Biagi, Metronidazole in the treatment of balantidiasis, Trans. Roy. Soc. Trop. Med. Hyg., 67:143 (1973).
14. D. R. Botero, Effectiveness of nitrimidazine in treatment of Balantidium coli infections, Trans. Roy. Soc. Trop. Med. Hyg., 67:145 (1973).
15. R. Muller, Guinea worm disease: epidemiology, control and treatment, Bull. W.H.O., 57:683 (1979).
16. L. Lasagna, The development and regulation of new medications. Science, 200:871 (1978).
17. I. de Carneri, The place of chemotherapy in the control of parasitoses, Proceedings of the I Mediterranean Congress of Chemotherapy, Madrid, 18-20 September 1978. p. 1003.
18. F. W. Jennings, G. M. Urquhart, P. K. Murray and B. M. Miller, Berenil and nitroimidazole combinations in the treatment of Trypanosoma brucei infection with central nervous system involvement, Int. J. Parasitol., 10:27 (1980).
19. A. C. Cuckler, C. M. Malanga and J. Conroy, Therapeutic efficacy of new nitroimidazoles for experimental trichomoniasis, amebiasis and trypanosomiasis, Am. J. Trop. Med. Hyg., 19:916 (1970).
20. E. Winkelmann and W. Raether, New chemotherapeutically active nitroimidazoles, 19th Interscience Conference on Antimicrobial agents and chemotherapy, Boston, Mass., 1-5 October 1979, Abstract 152.
21. E. W. J. de Maar, The future of tropical chemotherapy, Cronache Farmaceutische, 23:84 (1980).

22. The Medical Letter, Drugs for parasitic infections, 21:105 (December 28, No. 26) (1979).
23. A. Polak and R. Richle, Mode of action of the 2-nitroimidazole derivative benznidazole, Ann. Trop. Med. Parasitol., 72:45 (1978).
24. B. P. Goldstein, R. R. Vidal-Plana, B. Cavalleri, L. Zerelli, G. Carniti and L. G. Silvestri, The mechanism of action of the nitro-heterocyclic antimicrobial drugs. Metabolic activation by micro-organisms, J. Gen. Microbiol., 100:283 (1977).
25. C. M. Beard, K. L. Noller, W. M. O'Fallon, L. T. Kurland and M. B. Dockerty, Lack of evidence for cancer due to use of metronidazole, New Engl. J. Med., 10:519 (1979).

RESISTANCE TO 5-NITROIMIDAZOLES IN PATHOGENIC MICROORGANISMS

Miklós Müller

The Rockefeller University
New York, N.Y.10021, USA

ABSTRACT

Resistance (lowered susceptibility) to metronidazole and other 5-nitroimidazoles occurs, albeit infrequently, in clinical isolates of the normally susceptible anaerobic microorganisms, *Trichomonas vaginalis* and *Bacteroides* spp. Such resistance has been induced by *in vivo* drug pressure in *T. vaginalis* and *Tritrichomonas foetus* and by *in vitro* drug pressure in *B. fragilis*. All organisms studied showed cross-resistance to other 5-nitroimidazoles expressed to a different degree. The detection of resistance depends on the assay method used. The mechanism of resistance in natural isolates and in strains of trichomonads developed *in vivo* is unknown but seems to be connected with aerobic conditions. Resistance of *in vitro* developed *T. foetus* strains and of all *B. fragilis* strains studied is related to a decreased activity of pyruvate:ferredoxin oxidoreductase, responsible for the metabolic activation of 5-nitroimidazoles.

INTRODUCTION

Microorganisms often develop lowered susceptibility (i.e. resistance) towards antimicrobial agents. Metronidazole and related 5-nitroimidazole derivatives have been in clinical use against infections due to anaerobic protozoa and bacteria for about twenty years. Several authors warned that 5-nitroimidazole resistance will emerge, especially since microorganisms with lowered susceptibility can be obtained in laboratory experiments relatively easily. In this paper we will review this question based on data selected from studies in which the microorganisms were characterized in some detail (Table 1). Most data refer to

Table 1. Strains of Microorganisms Characterized in this Review

Species	Strain	Origin	Ref.
Trichomonas vaginalis	ATCC 30001	Standard strain from the American Type Culture Collection (Rockville, MD).	1
		Random isolates used as controls	
	A	from Vienna, Austria.	1
	E		2
	NYH 209	from New York, NY.	3
	NYH 272		
	NYH 286		
	F	from Paris, France.	4
	Du		
	F 60	From F under *in vivo* metronidazole pressure.	4
	Du 45	From Du under *in vivo* metronidazole pressure.	
		Isolates from treatment failures	
	IR-78	from Vienna, Austria.	1,5
	BO	from Gottenborg, Sweden.	6
	Boston	from the named cities, MA & NY.	3
	Albany		
	Fall River		
Tritrichomonas foetus	KV_1	Clone of isolate from Karlovy Vary, Czechoslovakia.	
	KV_1-M100	From KV_1 under *in vivo* (mouse) metronidazole pressure.	7
	KV_1-MR-100a	From KV_1 under *in vitro* metronidazole pressure. Highest drug concentration used was 100 μg/ml.	8
	KV_{c1}-MR-2	From a clone of KV_1 as above. Highest concentration used was 2 μg/ml.	8
Bacteroides fragilis	AM 17 AM 78	Wild type isolates.	9
	AM 75 AM 78.5	Metronidazole resistant derivatives of AM 17 and AM 78 obtained through mutagenesis and selection under drug pressure.	9

metronidazole. In view of the fundamental similarities in the mode of action of 5-nitroimidazoles (10) used in anaerobic infections, the general conclusions drawn are likely to be applicable to all these compounds. Honigberg analysed in detail the drug resistance in trichomonads giving an extensive bibliography through 1977 (11), thus no comprehensive coverage is attempted here. Therapeutic and other clinical studies, although of great importance, will not be discussed. We will concentrate on Trichomonas vaginalis, for comparative purposes on Tritrichomonas foetus, a parasite of cattle and on Bacteroides fragilis. Although information on other anaerobic protozoa (Entamoeba histolytica and Giardia lamblia) would be of great importance, practically no data are available on this topic.

Abbreviations used in this review are as follows: CD_{50} and CD_{95}: oral doses necessary to cure 50% or 95% of the infected animals. Mice were used in all tests quoted. CD(d): dose in mg drug per kg body weight per day; CD(t): single dose administered three times (2, 16 and 24 hours after infection) in mg drug per kg body weight. MLC and MIC: minimum lethal concentration and minimum inhibitory concentration in µg per ml. MLC_{24} and MLC_{48}: MLC values determined after 24 hr or 48 hr incubation, respectively.

DETERMINATION OF 5-NITROIMIDAZOLE SUSCEPTIBILITY

Resistance of certain strains of a protozoan or bacterial species normally susceptible to an antimicrobial compound is usually a relative property, i.e. a susceptibility which is lower, often by one or more orders of magnitude, than that of the majority of the strains. If development of such strains was achieved or observed under controlled conditions, when the initial strain remains available for comparison, then even relatively small changes in susceptibility can be detected and studied. In surveys of clinical isolates of a given species, however, the observed susceptibility levels will show a considerable variation and any definition of resistance will include a significant pragmatic element, i.e. the ease and safety of achieving in the host a drug level higher than that necessary to inhibit or kill the microorganism. These self-evident points are emphasized here because they help to clarify certain conflicting issues in the literature.

Results of in vitro assays are not always easy to interpret, thus, in vivo assays can be of great importance. In studies on 5-nitroimidazole resistance of trichomonads subcutaneous or intraperitoneal inoculation of the pathogen into mice, followed by various schedules of drug administration was used most frequently (4,7,12,13,14). Intravaginal infections are difficult to obtain (15). Absence of microorganisms at the site of inoculation is scored as cure and the results are expressed as CD_{50} or CD_{95} values. Interestingly, different doses are necessary to cure the subcutaneous and intraperitoneal infections (1,2). In vivo assays were not used to test 5-nitroimidazole resistance of B. fragilis.

Methods used to determine the in vitro nitroimidazole suseeptbility of microorganisms vary from laboratory to laboratory and results reported by different groups are difficult to compare. Of great significance are the following factors:

(a) The composition of the gas phase in the assay system. Aerobiosis strongly inhibits the antimicrobial action of 5-nitroimidazoles possibly because oxygen competes with the reductive activation (10). Poor anaerobiosis in the assay has been reported to give unexpectedly low susceptibility values for anaerobic bacteria (16). In the study of 5-nitroimidazole resistance of trichomonads, however, aerobic assays (1,3,17) proved to be of great value, as discussed below. Such assays are performed in plastic microtiter plates, the wells of which contain 1.5 ml of less of medium, kept in aerobic incubators. The medium equilibrates with air and thus the conditions are essentially aerobic. The conditions in this assay are, however, suboptimal for the trichomonads. No MIC but only MLC values can be obtained for most T. vaginalis strains because they do not multiply under aerobic conditions.

(b) Size of the inoculum. The antimicrobial activity of the 5-nitroimidazoles depends on their metabolic activation producing short-lived toxic intermediates and biologically inactive end-products. If susceptible organisms are present at high numbers in a medium containing a 5-nitroimidazole in a relatively low concentration, the drug will be rapidly converted to inactive products and its concentration in the assay system will not be constant (18,19).

(c) Composition of the medium used in the assay. This effect has been observed often but has not been explored in a systematic way. It is clear, however, that the presence or absence of certain components of culture media can markedly affect the 5-nitroimidazole susceptibility of microorganisms. The effect of ascorbate on T. foetus will be quoted as an example.

It is not the aim of this review to discuss the pharmacokinetics of 5-nitroimidazoles (20-22). It is to be mentioned only that the peak serum levels of the order of 20 $\mu g.ml^{-1}$ or above can easily be reached with standard single dose (2g) therapy with metronidazole and with other 5-nitroimidazoles. Peak levels increase proportionally with increasing doses administered. Tissue levels are usually similar to levels determined in serum. On this basis, microorganisms showing in vitro MIC values lower than 20 $\mu g.ml^{-1}$ are usually regarded as susceptible to the drugs in question. In vivo CD values obtained in mice cannot be directly extrapolated to humans because of different pharmacokinetic behaviour of 5-nitroimidazoles in various mammals. CE_{95} (t) values of less than 35

$mg.kg^{-1}$ can be regarded as reflecting normal susceptibility in *T. vaginalis*. No similar estimates are available for the other microorganisms discussed.

CLINICAL ISOLATES WITH LOWERED 5-NITROIMIDAZOLE RESISTANCE

The clinical efficacy of 5-nitroimidazoles in infections caused by anaerobic protozoa and bacteria is consistently high, suggesting that only few strains of the susceptible microorganisms are resistant to these drugs. Isolates with lowered susceptibility were detected, however.

Several *T. vaginalis* isolates obtained from persistent cases and from cases refractory to treatment were found to show lowered metronidazole susceptibility in *in vivo* mouse assays (Table 2) (1, 3,5). The demonstration of more than an order of magnitude higher

Table 2. Metronidazole Susceptibility of *Trichomonas vaginalis* isolates[a]

	In vivo test in mice	*In vitro* test aerobic	*In vitro* test anaerobic
	CD_{95}(t)	MLC_{24}	
SUSCEPTIBLE			
ATCC 30001	not infective	2	0.5
NYH 209	9.0	25	1
NYH 272	6.2	6	1
NYH 286	8.5	10	0.5
A	15	5 (3.1)[b]	1 (0.4)
Random isolates (n=70)	>35[c]		
RESISTANT			
IR-78	>>800	150 (100)	4 (0.4)
BO[d]		100	5
Boston	103	75	5
Albany	81	60	4
Fall River	150	250	2

[a]From reference (3).
[b]MLC values in brackets from reference (1).
[c]From reference (2).
[d]Data from author's laboratory on an isolate from case described in reference (6).

CD values for these isolates than for the control strains strongly suggests that the lowered susceptibility of the protozoon contributed to the therapeutic failure. Increased doses of metronidazole proved to be curative in the patients in question, supporting the above conclusion.

One of the above isolates studied in more detail showed lowered in vivo susceptibility not only to metronidazole but also to tinidazole and nimorazole (Table 3) (1). The degree of resistance was highest for metronidazole, intermediate for tinidazole and lowest for nimorazole. Comparative studies on a large number of similar isolates and with additional nitroimidazoles are needed before any therapeutic implications can be drawn.

Table 3. Efficacy of 5-nitroimidazoles on Subcutaneous <u>Trichomonas vaginalis</u> infections in mice

	Strain: Susceptible		Strain: Resistant		Relative resistance (Resistant/susceptible)	
DEVELOPED UNDER METRONIDAZOLE PRESSURE[a]						
	F	CD_{50} (d)	F60			
Metronidazole	8.5		460		59	
Nimorazole	11		450		26	
DEVELOPED UNDER NIMORAZOLE PRESSURE[a]						
	Du	CD_{50} (d)	Du45			
Metronidazole	7.5		500		67	
Nimorazole	15		500		34	
FRESH ISOLATES[b]						
	E	CD_{50} (t)	IR-78			
Metronidazole	11	(15)[c]	>800	(122)	>72	(8.1)
Nimorazole	40	(48)	146	(90)	3.7	(1.9)
Tinidazole	10	(16)	136	(74)	14	(4.6)

[a] From reference (4)

[b] From reference (1)

[c] Data in brackets represent values from intraperitoneal infections.

In vitro assays present serious difficulties in the assessment of 5-nitroimidazole susceptibility of trichomonas isolates. Assays performed under anaerobic conditions show MLC and MIC values falling within a relatively wide range (Table 2) (3) which, however, are almost entirely within the range regarded to reflect normal susceptibility. Thus *T. vaginalis* strains, whether susceptible or resistant to metronidazole *in vivo*, will be regarded as susceptible if assayed anaerobically.

MLC values obtained in the aerobic assay permitted the classification of the isolates into susceptible and resistant groups in a small study (3). There was a clear qualitative correlation between the assignment made on the basis of the aerobic *in vitro* assay and that made on the basis of the mouse assay. The values obtained, however, cannot be converted to drug blood levels necessary to eliminate the infection in the patient. Thus at present the only usefulness of the assay is that it detects strains with lowered susceptibility. Further detailed studies on a significant number of strains are needed to define the correlation between *in vitro* MLC values obtained with various methods and *in vivo* susceptibility. Although the aerobic assay seems to provide clearer indication of lowered *in vivo* susceptibility in *T. vaginalis*, probably special attention should be paid to the anaerobic assays in which certain strains give MIC values, although within the "susceptible" range, but significantly higher than the average values.

The frequency of 5-nitromidazole resistance in *T. vaginalis* cannot be assessed yet. The resistant strains discussed above were isolated from selected cases. A survey performed with *in vivo* and *in vitro* assays in Vienna detected no resistance among 94 fresh isolates (2). The question can be answered only through an extensive survey of random populations and through a careful study of treatment failures.

There are no conclusive experimental studies on the occurrence of 5-nitroimidazole resistance in other anaerobic protozoa. Since 5-nitroimidazoles are extensively used in the treatment of infections due to such microorganisms (*E. histolytica* and *G. lamblia*) studies on treatment failures would be highly desirable. Suitable assay methods were developed recently for *E. histolytica* (23). *G. lambia*, an organism notoriously difficult to isolate, can be assayed only on samples obtained directly from patients (24).

5-nitroimidazole resistance in clinical isolates of normally susceptible anaerobic bacteria has been reported only in exceptional cases. High MIC values were reported for a *B. fragilis* strain isolated from the colon of a patient undergoing long term metronidazole therapy for Crohn's disease (70 μg/ml) (25), for another from a urinary tract infection (150 μg/ml) (9), and for a *B.*

distasonis strain isolated from the peritoneum of an untreated patient (64 µg/ml) (26). The lowered susceptibility remained stable on subcultures. The resistance of the first of these strains could not be transferred and proved to be not mediated by a plasmid. The frequency of such strains is unknown but must be low since extensive screening programs for antibiotic susceptibility detected only few resistant strains.

IN VIVO DEVELOPMENT OF RESISTANCE

Several authors succeeded in increasing the 5-nitroimidazole tolerance of *T. vaginalis* and *T. foetus* by serious infection and subcurative treatment of laboratory animals (mice, hamster, etc.) (7,11-14). The development of resistance required several passages in each case and when studied showed a gradual increase. The increase varied from study to study. In *T. vaginalis* the maximum increase observed was thirtyfold (13) and in *T. foetus* the highest reported relative resistance exceeds 68 (7).

Lowered susceptibility to metronidazole is accompanied by lowered susceptibility to other 5-nitroimidazoles (Table 4) (7). Cross resistance in *T. vaginalis* was also demonstrated by exposing one strain to metronidazole and another to nimorazole (4). The in vivo susceptibilities to either drug changed to the same extent in both strains (Table 3).

Table 4. Efficacy of 5-Nitroimidazoles on Intraperitoneal *Tritrichomonas foetus* Infections in Mice[a]

Substance	Strain		Relative resistance
	KV_1	KV_1-M100[a]	(KV_1-M100/KV_1)
	CD_{50}(t)		
Metronidazole	11.7	> 800	> 68
Nimorazole	24.3	> 800	> 33
Tinidazole	3.5	140	40
Ornidazole	6.5	180	28
Panidazole	7.9	137.5	17
Flunidazole	12.1	302	25
MK 910	2.5	80	32
Ronidazole	1.7	42	25
Dimetridazole	7.5	200	27

[a]From reference (7).

In vitro testing of the various resistant strains gave similar results as obtained in clinical isolates. With standard assays most strains exhibited MIC or MLC values regarded as reflecting susceptibility (4,11-14). An anlysis of the susceptibility pattern of the *in vivo* developed resistant *T. foetus* strain revealed that standard assays performed in test tubes and in microtiter plates under anaerobic conditions give similar values for this strain and its susceptible parent. Under aerobic conditions, however, the resistant strain showed high MLC values (Table 5)(17). The selection of the medium was of importance, too. In media containing ascorbic acid, no or little difference was seen between the two strains (17).

These results gave the impetus to test the aerobic assay and *T. vaginalis* too, as discussed before. Unfortunately, no similar *in vitro* data are available for *T. vaginalis* strains which developed resistance under *in vivo* drug pressure.

Table 5. Metronidazole Susceptibility in vitro (tray test) of *Tritrichomonas foetus*[a]

Medium	KV_1		KV_1-M100	
	aerobic	anaerobic	aerobic	anaerobic
	MLC_{48}		MCC_{48}	
Fluid thioglycolate	1.6	1.5	100	1.0
+ 0.1% ascorbic acid	1.6		25	
CACH (Müller & Gottschalk)	0.8		0.8	
- ascorbic acid	3.1		50	
TYM (Diamond)	1.6	1.0	50	1.5
TTYS (Diamond)	1.6		100	
+ 0.1% ascorbic acid	0.4		1.6	

[a]From reference (17).

IN VITRO DEVELOPMENT OF RESISTANCE

Development of resistance was achieved in *T. foetus* and *T.* vaginalis by exposing the organisms in culture to sublethal concentrations of metronidazole over a large number of subcultures. Relative resistance in *T. vaginalis* was usually moderate (14,27) in T. foetus; however, spectacular results were obtained (Table 6) (8). Authors assume that resistance arises by mutation elicited by metronidazole, and is then selected for by the presence of the drug.

In *B. fragilis* other mutagens were used before selection with metronidazole was applied (Table 6) (9). The data do not permit the determination whether the frequency of appearance of resistance is compatible with rates expected for mutations. Irrespective of the mode of appearance, several strains showing high *in vitro* resistance were obtained. The resistance is stable and usually is not lost during cultivation in absence of drug.

MECHANISMS OF RESISTANCE TO 5-NITROIMIDAZOLES

Most organisms are but little affected by metronidazole and other 5-nitroimidazoles (10). The absence of susceptibility in these organisms is connected with two factors: (a) their metabolism is characterized by the absence of enzymatic systems (or more precisely of electron carriers) which can reductively activate the 5-nitroimidazoles with great efficiency; and (b) they live under aerobic conditions, mentioned above as being greatly inhibitory to 5-nitroimidazole action. As an oversimplification, an anaerobic organism which developed lowered 5-nitroimidazole susceptibility is somewhat "less anaerobic" than its wild type predecessor.

Table 6. In vitro Development of Metronidazole Resistance

Parent organism		MIC	Treatment	Derivative	MIC	Ref.
Tritrichomonas foetus KV_1		1.0	Increasing metronidazole concentration up to 100 µg/ml	KV_1-MR100a	400	8
Bacteroides fragilis	AM 17	1.0	Mutagenesis and selection on plates with metronidazole at 25 µg/ml	AM 75	100	9
	AM 78	2.5	10 µg/ml	AM 78.5	25	

The microbicidal activity of 5-nitroimidazoles is related to the efficient reduction of the nitro-group by electrons generated in certain key steps of the energy metabolism (10). In trichomonads and in B. fragilis this step is the oxidative decarboxylation of pyruvate by pyruvate:ferredox in oxidoreductase leading via additional steps to the formation of acetate. Ferredoxin and similar low redox potential electron transport proteins function as the proximal reductant for the nitro group of the drugs. In T. foetus strains made resistant in vitro and in highly resistant B. fragilis the activity of this enzyme is not detectable or several orders of magnitude lower than in the parent strains (Table 7) (8, 9,28). Decreased metronidazole reduction and low nitroreductase activity, probably reflecting the same process, was observed in one of the clinical B. fragilis isolates (29). In moderately resistant strains of both organisms the activity is significantly lower than in the parent strains. In both the protozoon and the bacterium the pattern of eliminated end products of carbohydrate metabolism changes (8,30). In agreement with the enzyme deficiency less or no acetate is produced and much of the carbon which would end up as acetate is eliminated as ethanol. Anaerobic activation of metronidazole is markedly decreased in the resistant B. fragilis (29). Although other enzyme changes are also observed in resistant T. foetus, their significance in the mechanism of resistance remains to be elucidated.

In contrast to the simple enzyme deficiency possibly underlying the resistance apparent in vitro under anaerobic conditions, no such changes could be detected yet in the T. vaginalis strains that show significant resistance only under aerobic conditions. Pyruvate: ferredoxin oxidoreductase shows some strain dependent variation in T. vaginalis but susceptible and resistant strains show no systematic differences in the activity of this enzyme (Table 7) nor in the elimination of acetate (31). Thus an intact low redox potential system is present and the anaerobic activation of metronidazole (as measured by the intracellular accumulation of [^{14}C]metronidazole (32) by resistant trichomonads is similar to that in susceptible strains (Table 8). In view of the marked effect of aerobiosis observed in resistant trichomonads, the respiration and the activity of the major terminal oxidase, the cytoplasmic NADH oxidase has been determined (Table 7). Again the small variations observed could not be correlated with the level of susceptibility.

The only difference between the resistant and susceptible T. vaginalis strains is that aerobiosis inhibits the metabolic activation of metronidazole to a significantly greater degree in the less susceptible strains (Table 8) (31). Although this observation might explain the observed aerobic resistance, the molecular mechanisms underlying this effect remain obscure.

Table 7. Enzyme Activities in Anaerobic Microorganisms

	Pyruvate:ferredoxin oxidoreductase	NADH oxidase	Catalase[a]
	nmol . min^{-1} . mg^{-1} protein		
Tritrichomonas foetus[b]			
SUSCEPTIBLE			
KV_1	273	576	340
RESISTANT			
KV_1-MR-100a	<0.1	1.1	0.44
KV_{c1}-MR-2	6.4	1.3	608
Trichomonas vaginalis[c]			
SUSCEPTIBLE			
ATCC 30001	106	20	
NYH 209	121	370	
NYH 272	121	550	
NYH 286	85	380	
RESISTANT			
IR-78	97	830	
Boston	108	730	
Albany	111	310	
Fall River	139	410	
Bacteroides fragilis[d]			
SUSCEPTIBLE			
AM 17	91[a]		
AM 78	120		
RESISTANT			
AM 75	<0.1		
AM 78.5	5.1		

[a]Arbitrary units

[b]From reference (8)

[c]From reference (31)

[d]From reference (9)

Table 8. Uptake of [^{14}C]metronidazole by <u>Trichomonas vaginalis</u> isolates[a]

	Anaerobic conditions	Aerobic conditions	Anaerobic / Aerobic
	μg drug . 15 min^{-1} . mg^{-1} protein		
SUSCEPTIBLE			
ATCC 30001	4.26	0.42	10
NYH 209	3.00	0.26	11
NYH 272	3.75	0.40	9
NYH 286	3.63	0.23	16
A	1.86	0.12	15
RESISTANT			
IR-78	3.22	0.05	65
Boston	1.70	0.07	26
Fall River	2:02	0.05	42

[a]From reference (31).

CONCLUSIONS

The findings summarized above demonstrate that organisms normally susceptible to antimicrobial 5-nitroimidazoles can acquire lowered susceptibility. The development of resistance can be achieved in laboratory experiments relatively easily. Resistant strains of <u>Bacteroides</u> spp. and of <u>T. vaginalis</u> were isolated from patients as well. The frequency of occurrence of such resistant strains is not known but the relatively unchanged clinical efficacy of metronidazole during the past two decades suggests that it is low. A detailed survey of metronidazole susceptibility of clinical isolates with appropriate test methods seems to be warranted for <u>T. vaginalis</u>, and, whenever suitable methods become available, for other susceptible protozoa. Since a 5-nitroimidazole is usually included in susceptibility surveys of anaerobic bacteria, data accumulate rapidly.

The clinical significance of 5-nitroimidazole resistance is not known. In human trichomoniasis treatment failures were usually attributed to other causes in the past (33,34). Only further studies can establish the relative contribution of drug resistance and other factors to therapeutic difficulties observed in different populations. In the absence of non-nitroimidazole drugs effective in trichomoniasis, this is an important question.

The availability of several antibiotics against anaerobic bacterial infections makes the question of the 5-nitroimidazole resistance in various Bacteroides species less urgent, though increasing use of these drugs might reveal more cases of therapeutic problems.

Studies on some 5-nitroimidazole resistant microorganisms provide interesting data supporting our current understanding of the mode of antianaerobic action of 5-nitroimidazoles, and should give further insight into this important question.

ACKNOWLEDGEMENTS

Generous advice on the practical and theoretical aspects of 5-nitroimidazole resistance by Dr. J.G. Meingassner (Vienna, Austria) and by Drs. A. Čerkasovovà and J. Čerkasov (Prague, Czechoslovakia) is gratefully acknowledged. The latter also kindly permitted the use of "in press" information from their laboratory. Thanks are due to D.G. Lindmark and T.E. Gorrell for discussions and for performing many of the experiments (published and unpublished) on which this presentation is based, to E. Wolff, C. DiDelmonico, and D. Fishel for enthusiastic and competent technical help, and to M.W. Morse for work on the manuscript. Original research in author's laboratory was supported by U.S. Public Health service grants AI 11942 and AI 12932 from the National Institute of Allergy and Infectious Diseases and from Searle Laboratories, Division of G.D. Searle & Co., Chicago, Illinois.

REFERENCES

1. J. G. Meingassner and J. Thurner, Strain of Trichomonas vaginalis resistant to metronidazole and other 5-nitro-imidazoles, Antimicrob. Agents Chemother. 15:254 (1979).
2. J. G. Meingassner, L. Havelec, and H. Mieth, Studies on strain sensitivity of Trichomonas vaginalis to metronidazole, Brit. J. Vener. Dis. 54:72 (1978).
3. M. Müller, J. G. Meingassner, W. A. Miller and W. J. Ledger, Three metronidazole-resistant strains of Trichomonas vaginalis from the USA, Am. J. Obst. Gyn. (in press).
4. F. Benazet and L. Guillaume, Induction of in vivo resistance of Trichomonas vaginalis to nitrimidazine, Lancet 2:982 (1971).
5. J. Thurner and J. G. Meingassner, Isolation of Trichomonas vaginalis resistant to metronidazole, Lancet 2:738 (1978).
6. A. Forsgren and L. Forssman, Metronidazole-resistant Trichomonas vaginalis, Brit. J. Vener. Dis. 55:351 (1979).
7. J. G. Meingassner and H. Mieth, Cross-resistance of trichomonads to 5-nitroimidazole-derivatives, Experientia 23:183 (1976).
8. A. Čerkasovovà, J. Čerkasov, J. Kulda and P.Demes, Metronidazole action on Tritrichomonas foetus: enzyme activities in the strains of a different resistance to the drug in: "The Host-Invader

Interplay. Third International Symposium on the Physiology of Parasites and Host-Parasite Relationships", H. van den Bossche, ed., Elsevier/North-Holland Biomedical Press, Amsterdam (1980).
9. M. L. Britz and R. G. Wilkinson, Isolation and properties of metronidazole-resistant mutants of Bacteroides fragilis, Antimicrob. Agents Chemother. 16:19 (1979).
10. M. Müller, Mode of action of metronidazole on anaerobic microorganisms, in: "Metronidazole: Proceedings of the 2nd International Symposium on Anaerobic Infections, Series No.18", I. Phillips and J. Collier, eds., The Royal Society of Medicine, London (1979).
11. B.M. Honigberg, Trichomonads of importance in human medicine, in: "Parasitic Protozoa, Vol.II", J.P. Kreier, ed., Academic Press, New York (1978).
12. I. de Carneri, Perspectives in the treatment of protozoal diseases resistant to metronidazole, Trans. Roy. Soc. Trop. Med. Hyg. 65:268 (1971).
13. I. de Carneri and F. Trane, In vivo resistance to metronidazole induced on four recently isolated strains of Trichomonas vaginalis, Drug Res. 21:377 (1971).
14. P. Durel, J. Couture and M.T. Bassoullet, Commentaires sur deux cas de vaginite à T. vaginalis guéris avec difficulté par le métronidazole, Wiad. Parazytol. 12:422 (1966).
15. P. Actor, D.S. Ziv and J. F. Pagano, Resistance to metronidazole by Tritrichomonas foetus in hamsters infected intravaginally, Science 164:439 (1969).
16. E. Shelagh, E. Milne, J. Stokes and P.M. Waterworth, Incomplete anaerobiosis as a cause of metronidazole 'resistance'. J. Clin. Pathol. 31:933 (1978).
17. J. G. Meingassner, H. Mieth, R. Czok, D.G. Lindmark and M. Müller, Assay conditions and the demonstration of nitroimidazole resistance to Tritrichomonas foetus. Antimicrob. Agents Chemother. 13:1 (1978).
18. E. D. Ralph and D. A. Clarke, Inactivation of metronidazole by anaerobic and aerobic bacteria, Antimicrob. Agents Chemother. 14:377 (1978).
19. B. Korner and H. K. Jensen, Sensitivity of Trichomonas vaginalis to metronidazole, tinidazole, and nifuratel in vitro, Brit. J. Vener. Dis. 52:404 (1976).
20. G. W. Houghton, P. S. Thorne, J. Smith, R. Templeton and J. Collier, Comparison of the pharmacokinetics of metronidazole in healthy female volunteers following either a single oral or intravenous dose, Br. J. Clin. Pharmac. 8:337 (1979).
21. E. D. Ralph and M. M. Kirby, Human pharmacology of oral metronidazole as determined by bioassay, in: "Metronidazole: Proceedings of the International Metronidazole Conference, Montreal, Quebec, Canada", S. M. Finegold, ed., Excerpta Medica, Amsterdam (1977).

22. R. Templeton, Metabolism and pharmacokinetics of metronidazole: a review, in "Metronidazole: Proceedings of the International Metronidazole Conference, Montreal, Quebec, Canada", S.M. Finegold, ed., Excerpta Medica, Amsterdam (1977).
23. F. D. Gillin and L. S. Diamond, Clonal growth of Entamoeba histolytica in agar, J. Protozool. 41:317 (1980).
24. L. Jokipii and A. M. Jokipii, In vitro susceptibility of Giardia lamblia trophozoites to metronidazole and tinidazole, J. Infect. Dis. 141:317 (1980).
25. H. R. Ingram, S. Eaton, C. W. Venables and P. C. Adams, Bacteroides fragilis resistant to metronidazole after long term therapy, Lancet 1:214 (1978).
26. V. O. Rotimi, B. I. Duerden, V. Ede and A. E. MacKinnon, Metronidazole-resistant bacteroides from untreated patient, Lancet 1:833 (1979).
27. B. M. Honigberg and M. C. Livingston, In vitro sensitivity of Trichomonas vaginalis to metronidazole, Proc. First Int. Congr. Parasitol. 1:365 (1966).
28. A. Čerkasovovà and J. Čerkasov, Absence of pyruvate splitting processes in Tritrichomonas foetus resistant to metronidazole, J. Protozool. 25:34A (1978).
29. F. P. Tally, D. R. Snydman, M. J. Shimell and B. R. Goldin, Mechanisms of antimicrobial resistance of Bacteroides fragilis, in: "Metronidazole: Royal Society of Medicine International Congress and Symposium Series No.18", I. Phillips and J. Collier, eds., The Royal Society of Medicine, London (1979).
30. A. Čerkasovovà, J. Čerkasov and J. Kulda, Ethanol formation in the cell of Tritrichomonas foetus, J. Protozool. 26:39A (1979).
31. M. Müller, Physiological studies on Trichomonas vaginalis with lowered susceptibility to metronidazole, J. Protozool. 27:26A (1980).
32. M. Müller and D. G. Lindmark, Uptake of metronidazole and its effect on viability in trichomonads and Entamoeba invadens under anaerobic and aerobic conditions, Antimicrob. Agents Chemother. 9:696 (1976).
33. J. A. McFadzean, I. M. Pugh, S. L. Squires, and J.P.F. Whelan, Further observations on strain sensitivity of Trichomonas vaginalis to metronidazole, Brit. J. Vener. Dis. 45:161 (1969).
34. D. I. Edwards, E. J. Thompson, J. Tomusange, and D. Hanson, Inactivation of metronidazole by aerobic organisms, J. Antimicrob. Chemother. 5:315 (1979).

PHARMACOKINETICS OF NITROIMIDAZOLES

T.R. Marten, R.J. Ruane, J.A. White and L. Clarke

Roche Products Ltd.

Welwyn Garden City, England

In this paper I want to try and show how a knowledge of the metabolism and pharmacokinetics of a compound helps us to understand the reasons why it may or may not satisfy the criteria desired of a particular drug. How we want a drug to behave depends on the use we have in mind for it. For example, in Fig. 1, are shown the plasma profiles for two nitroimidazole drugs which are used for the treatment of Trichomonas vaginalis. In this indication it is important to maintain plasma levels above the minimum inhibitory concentration (MIC) for as long a time as possible. It can be seen from the Figure that this level will be maintained longer when the compound with the longer elimination half-life, the time taken for the plasma concentration to drop to half its value, is administered. There is also the advantage that the peak levels are approximately equal for both compounds so the therapeutic ratio for ornidazole should be better than for metronidazole.

For a compound to be a successful radiosensitizer, one would like completely the opposite type of profile. Fig. 2 shows the optimum picture which one would like to see: a rapidly achieved peak level of the drug, which, ideally, would be in equilibrium with the tumour, followed by removal of the compound from the body as quickly as possible. This ensures high levels of the sensitizer, and hence a good enhancement ratio, and a short half-life which minimises the exposure and hence the side effects. However, it is almost impossible to achieve this type of profile from an orally administered drug and one is more likely to obtain the type of plasma picture shown in Fig. 1.

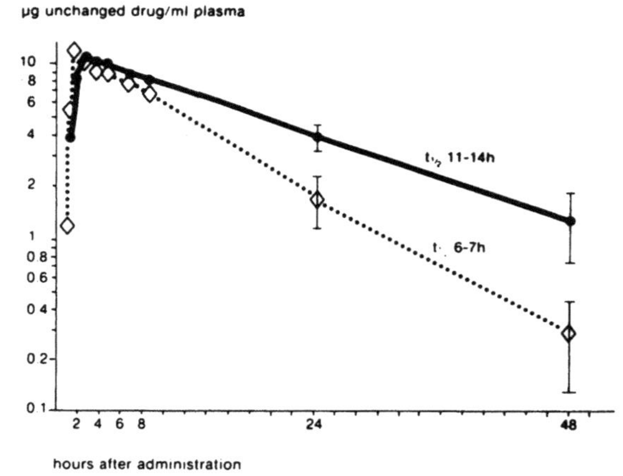

Fig. 1. Plasma levels of unchanged drug measured in the plasma of volunteers given 750 mg doses of ornidazole (●——●) or metronidazole (◇......◇) orally.

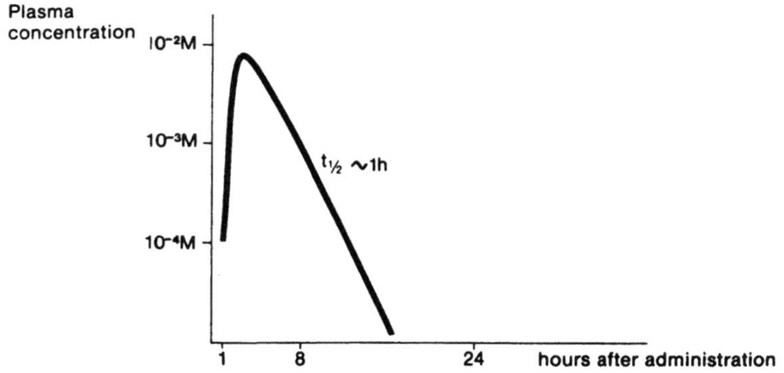

Fig. 2. Most desirable profile for a radiosensitizer.

I want to analyse the pharmacokinetics of misonidazole (Fig. 3) in humans, explain how the body copes with the compound and then show how we can use this information to choose an improved radiosensitizer.

Fig. 4 shows the type of plasma profile that is seen when misonidazole is given to humans. Firstly, we see that we need to give high doses to achieve levels of the drug in blood and in the tumour, where levels comparable to those reached in plasma have been found, in order to give a marked sensitizing effect. In theory, of course, we could carry on dosing until we reached higher levels, but this is not possible in practice because of the side effects of the drug. Secondly, although we get dose-related peak plasma levels of the compound, the half-life shows a tendency

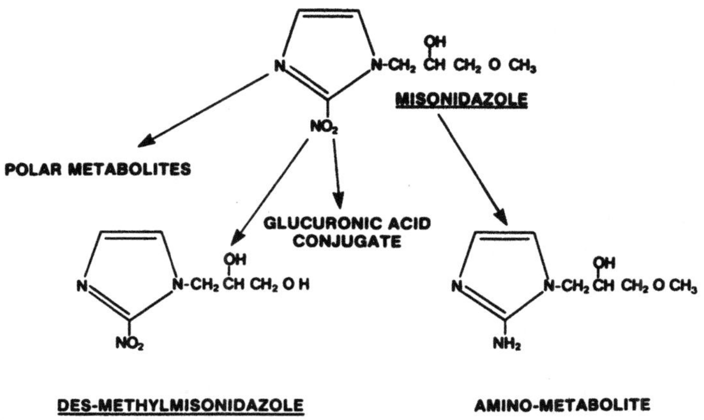

Fig. 3. Metabolic pathways of misonidazole.

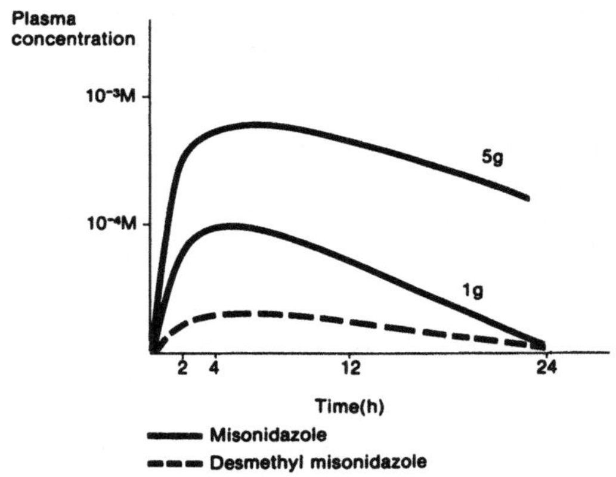

Fig. 4. Plasma levels following the oral administration of misonidazole.

to increase with increasing dose. This, coupled with the fact that there is a metabolite, desmethylmisonidazole (Ro 05-9963) (Fig. 3), present in the plasma at low and fairly constant levels, suggests that there is some relatively slow process which controls the elimination of misonidazole. We have shown in rats, and others have shown in mice (1), that this step is the O-demethylation of misonidazole to give desmethylmisonidazole.

Table 1 shows some of the pharmacokinetic parameters that may be derived following the intravenous administration of misonidazole to volunteers. The volume of distribution is a value which gives some indication of the volume throughout which the drug is distributed, and suggests that misonidazole is distributed throughout total body water, not unexpected for a compound which is non-protein bound and is of medium polarity (octanol:water partition

Table 1. Pharmacokinetic parameters of misonidazole in man

Half-life	10 - 14 hours
Volume of Distribution	40 - 50 litres
Plasma Clearance	40 - 70 ml/min
Percentage of dose excreted unchanged in the urine	15 - 20%
Renal Clearance	7 - 15 ml/min

coefficient = 0.4). The plasma clearance value is a description of the ability of the system to eliminate the drug and this relatively low value suggests that the elimination of the compound is not dependent on the flow of the blood through either of the two main organs of elimination, the kidney or the liver. We can see that relatively little of the administered compound finds its way into urine. As the compound is not significantly protein-bound, it will all be filtered out of the blood by the kidney glomerulus but the low renal clearance value indicates that the majority of the compound is reabsorbed in the tubule and does not reach the urine. The kidney is therefore not a particularly important organ for the elimination of the unchanged drug from the body.

The bioavailability, a measure of the amount of unchanged drug which reaches the circulation following an oral dose as compared to that following an intravenous dose, is high so the low amount of unchanged material excreted in the urine suggests that there must be other pathways of metabolism and indeed this is the case (Fig. 3). If we turn to the behaviour of the only metabolite detectable in the plasma, desmethylmisonidazole, a rather different picture is seen. Although much the same percentage of the dose is eliminated as this metabolite (Table 2), the renal clearance value is much higher. This shows that the kidney is much better able to eliminate this compound into the urine than the parent drug. In the case where the clearance approaches the glomerular filtration rate (∿125 ml/min) there is probably no re-absorption of the drug by the kidney.

Table 2. Pharmacokinetic parameters of desmethylmisonidazole derived from misonidazole

Percentage of misonidazole dose excreted in the urine as desmethylmisonidazole	20 - 25%
Renal clearance of desmethylmisonidazole	60 - 100 ml/min

This compound has been orally administered to several humans and the pharmacokinetic data are summarised in Table 3. Unfortunately we do not know what the bioavailability of desmethylmisonidazole is but a large proportion of the dose is eliminated unchanged in the urine and this is accompanied by a reasonably high renal clearance value. The plasma clearance value can only be quoted relative to the bioavailability but even if this is complete it is obvious that the kidney is an important organ for the elimination of this compound. Unlike misonidazole (Fig. 3) no glucuronide metabolite of this compound has been detected either in animals or man after receiving misonidazole or desmethylmisonidazole, indicating that the body does not have to increase the water solubility of this compound any further.

Table 3. Pharmacokinetic parameters of desmethylmisonidazole in man

Half-life	5 - 6 hours
Plasma clearance / Bioavailability	115 - 130 ml/min
Percentage of the dose eliminated unchanged in the urine	50 - 60%
Renal clearance	60 - 70 ml/min

The following summarises the results that have been found with misonidazole. The long half-life is due to reabsorption of the compound in the kidney and to a saturable metabolic pathway to desmethylmisonidazole. Although this compound is well cleared by the kidney, its rate of elimination in this instance is governed by its rate of formation from misonidazole and is therefore limited to a small proportion of the dose. The remaining misonidazole is eliminated by a number of pathways (Fig. 3). Although desmethylmisonidazole would appear to be a better compound on pharmacokinetic grounds, its intrinsic radiosensitization is no better than misonidazole and this probably mediates against its clinical use.

In order to choose a suitable successor to misonidazole a number of experiments in animals are carried out to compare new compounds with misonidazole. Table 4 outlines a number of the parameters, which were measured in humans and which have been measured in dogs both at Roche and by White et al. (3,4). As you will see, the overall pattern for misonidazole and desmethylmisonidazole is very similar to that in man. Data are also included for a compound, Ro 03-8799, which we are currently investigating. The salient pharmacokinetic features are its

Table 4. Pharmacokinetic parameters in the dog

	Misonidazole	Desmethyl-misonidazole	Ro 03-8799
Half-life (h)	4 - 5	2 - 3	1.5
Plasma Clearance (ml/min/kg)	1 - 2	2 - 3	27
Percentage excreted unchanged in the urine	5 - 7	75	13
Renal clearance (ml/min/kg)	0.08-0.09	2-2.5	3.4
Oral Bioavailability	High	Medium-High	Low

Structure of Ro 03-8799 and its major metabolite

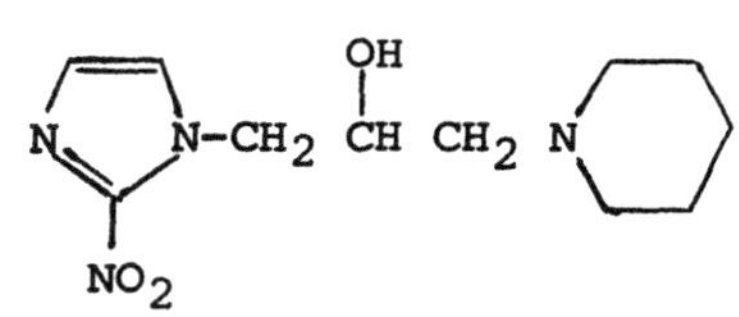

Ro 03-8799

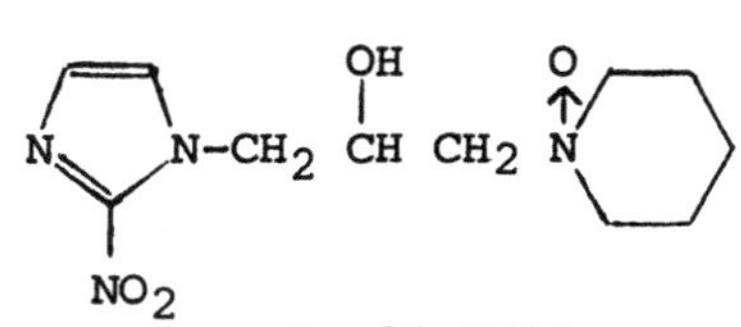

Ro 31-0313

short half-life and its high plasma clearance. It is interesting to note that the renal clearance is relatively low and, if we consider the extra-renal clearance value, we get a figure which approximates to a value of hepatic blood flow (25 ml/min/kg). This suggests that metabolism in the liver must be an important route of removal of the drug from the body and, indeed, a major metabolite of Ro 03-8799 is the N-oxide, Ro 31-0313. There is, however, one problem with this compound and that is its low bioavailability. It seems that because the compound is a good substrate for hepatic metabolism, there is a high first pass effect and little of the unchanged drug reaches the circulation. This, combined with slow absorption, which we have confirmed by in situ gut studies in rats, gives very low blood levels of the unchanged compound. However, because of the high radiosensitization afforded by this compound, we are considering developing this compound as an intravenous preparation so that we can achieve high initial levels, which we have already shown to give good tumour levels, and follow this up by rapid elimination. If we examine Fig. 2 we see that the desired profile has very nearly been reached. By abolishing the rapid absorption by giving the drug intravenously we achieve the profile that, by oral administration, would be almost impossible.

REFERENCES

1. P. Workman, Dose dependence and related studies on the pharmacokinetics of misonidazole and desmethylmisonidazole in mice, Cancer Chemotherapy and Pharmacology 5:27 (1980).
2. S. Dische, J. F. Fowler, M. I. Saunders, M.R.L. Stratford, P. Anderson, A. Minchinton, and M. Lee, A drug for improved radiosensitization in radiotherapy, Br. J. Cancer 42:153 (1980).
3. R.A.S. White, P. Workman, L.S. Freedman, L.N. Owen and N.M. Bleehen, The pharmacokinetics of misonidazole in the dog, Europ. J. Cancer 15:1233 (1979).
4. R.A.S. White and P. Workman, Pharmacokinetic and tumour-penetration properties of the hypoxic cell radiosensitizer desmethylmisonidazole (Ro 05-9963) in dogs, Br. J. Cancer 41:268 (1980).

THE CLINICAL PHARMACOLOGY OF MISONIDAZOLE IN RADIATION THERAPY ONCOLOGY GROUP TRIALS

Theodore L. Phillips and Todd H. Wasserman

Department of Radiation Oncology, University of California, San Francisco, and Department of Radiation Oncology, Washington University School of Medicine, St. Louis, Missouri
Supported by the Radiation Therapy Oncology Group CA21439

INTRODUCTION

Because the effectiveness of a hypoxic cell sensitizer of the imidazole class depends directly on its concentration in serum and within the tumor, the pharmacologic distribution of these drugs is of extreme importance. Numerous reports including those originally of Adams and colleagues (1), McNally (2) and of Denekamp (3) have indicated as well as have the skin experiments of Dische and colleagues (4), that the enhancement ratio observable is closely tied to the concentration in tumor and serum. The concentrations versus effectiveness curve for misonidazole rises steeply so that enhancement ratios between 1.3 and 1.5 may be observed at concentrations as low as 25-30 μg/ml within tumor. For these reasons, the determination of concentration within tumor and serum, a more readily-measurable quantity, has been extremely important in the Phase I and II studies of the RTOG (5-9).

METHODS OF DRUG LEVEL MEASUREMENT

As its inception, the RTOG Phase I evaluation of misonidazole employed both high performance liquid chromatography (HPLC) and ultraviolet spectrophotometric measurements of misonidazole concentration. Marques et al. at UCSF were first to describe the HPLC method for the determination of misonidazole and their metabolites in plasma and uring (10,5). Although both the ultraviolet spectrophotometric and HPLC methods have been employed in the RTOG, the HPLC method is favored since there are fewer pitfalls in its application and there is less error introduced by sample hemolysis. Both methods are outlined in detail in Tables I and 2

AVERAGE SERUM LEVELS IN PHASE I AND PHASE II RTOG TRIALS

Misonidazole is absorbed from the gastrointestinal tract after oral administration rather slowly. Although 90% of the peak level may be reached at approximately one hour, a plateau then ensues while further absorption occurs and this plateau may last from 4-6 hours after initial ingestion and is somewhat longer at higher total doses. Since the 4-6 hour period appears optimal for the administration of radiotherapy, it has been general practice in the RTOG trials to obtain serum samples in Phase I and II studies at the time of irradiation 4-6 hours after oral ingestion of the tablets or capsules. Table 3 shows the relationship between total ingested dose in g/m^2 and the mean 4-6 hour serum level with the standard deviation and observed ranges. Three-fourths of these determinations were done by HPLC and the figures quoted for levels include both misonidazole and desmethylmisonidazole (Ro 07-0582 and Ro 05-9963).

When these studies were expanded from the two initial institutions (5,8) (University of California at San Francisco and Roswell Park Memorial Institute, New York) to the entire RTOG (approximately 26 institutions) the majority of determinations shifted from HPLC to UV (9). It was then necessary to compare the serum mean 4-6 hour levels seen in the Phase I study to those seen in the Phase II study. This comparison is made in Table 4. It can be seen that levels were comparable in spite of the fact that levels were now being measured in 26 different laboratories rather than two initial laboratories of the Phase I study.

Mean 4-6 hour serum levels were 76 μg/ml for Phase I versus 68 μg/ml for Phase II at 2 g/m^2 and 46 μg/ml versus 44 μg/ml at 1.25 g/m^2, a very small difference.

In order to insure comparability of laboratories, the RTOG has employed a cross-calibration system. In this system, each member institution was required to set up and standardize its laboratory against pure drug obtained from the National Cancer Institute. The laboratory then requested a series of known and unknown samples from the standardization laboratory at UCSF, measured these levels and reported the results back to the standardization laboratory. If discrepancies were noted, consultations occurred and the methodology of the member laboratory was corrected. This system appears to have worked well since the vast majority of levels are in good agreement with the Phase I measurements done in only two institutions. A few exceptions to this have occurred, problems have been identified and corrected. The majority of level problems have occurred with ultraviolet spectrophotometry in which improper standards were utilized, hemolysis had occurred in samples and other pitfalls listed in Table I had occurred.

Forty-one measurements were made of the serum half-life in which samples were obtained every hour for the first few hours and then every four hours for the ensuing 24 hours on patients admitted to the Clinical Research Center at UCSF or to the Roswell Park Memorial Institute. The mean serum half-life was 13.8± 5.1 hrs (standard deviation) (7). In eight patients serial half-life determinations were available and there was no evidence of change in the half-life due to the induction of enzymes by the misonidazole (causing shortening of the half-life). There was a tendency to shorter half-lives with smaller total doses but in no case was this significant.

In patients in whom the major metabolite of misonidazole, desmethylmisonidazole, was measured using the HPLC method it was found to reach approximately 10% of the misonidazole level and thereafter parallel the misonidazole level. Since it is an equally effective sensitizer its concentration was combined with that of the misonidazole in the totals shown in the various tables.

The urinary excretion of misonidazole and desmethylmisonidazole was measured in 22 patients using 24-28 hour urinary samples with 4-hourly collections. An average of 30 ± 13% of the total administered drug was identifiable in the urine over the first 24 hours. Fifty-seven per cent of this material was desmethylmisonidazole and 43% misonidazole. Thus the excretion favored the metabolite although its concentration was only 10% of the parent compound of the serum. It is clear therefore that this more polar compound is more readily excreted by the kidneys and would necessarily have a shorter half-life if employed alone as the major sensitizer. It is also clear that a large portion of the parent compound is further broken down into fragments not detectable in the HPLC analysis. In a small number of patients tested, no detectable compound was found in the stool. This is not surprising since many bacterial species in the stool may break down misonidazole if it were to be excreted there by the liver.

Tumor levels were determined in three excised superficial tumor nodules and in all cases were found to be 80-90% of the peak serum level obtained simultaneously. One sample of pleural fluid was also examined six hours following oral ingestion of misonidazole and it exhibited 100% of the simultaneous serum level.

COMPARISON OF THE BIOAVAILABILITY OF MISONIDAZOLE IN TABLET AND CAPSULE FORM

After the first 102 patients were studied for misonidazole toxicity and pharmacology using enteric coated tablets, it was decided to switch the formulation to 500 and 100 mg capsules. In order to establish that this format gave similar blood levels, a second study of bioavailability was initiated which enlisted 43 patients (11). Patients were given various regimens and in addition, the administration of daily doses of 400-500 mg/m^2 was explored. This study also involved 9 children as well as 34 adults since the initial study had not involved testing for toxicity or pharmacology in children.

The mean 4-6 hour serum levels achieved with various dose schemes are shown in Table 5. For daily administration in adults it is interesting to note that 400 mg/m^2 gave levels of 15 and 16 μg/ml and that 500 mg/m^2 19 μg/ml. In both situations the levels measured on Friday after 5 daily doses were higher than on Mondays indicating some slight build-up of the compound with daily administration. With intermittend administration there was no evidence of any difference between the capsule and tablet form when larger numbers of patients were studied.

For children, daily misonidazole again gave somewhat higher levels on Fridays after 5 days of administration than after the first day. There was no significant difference between capsules and tablets in terms of the mean 4-6 hour blood level. Of the 34 adults, 25 had comparisons of tablets and capsules and of the 9 children, 5 had such comparisons as shown in Table 5.

Complete pharmacology curves were available in a limited number of patients receiving capsules and tablets as shown in Table 6. The mean half-life in adults was 12 ± 3 hours for capsules and 13.8 ± 4.8 for tablets. In children, the corresponding values are 10.9 ± 5.3 for capsules and 9.6 ± 5.5 (standard deviation) for tablets. Thus, there is no evidence of any difference in the half-life or the mean 4-6 hour serum concentration for capsules versus tablets and it is clearly safe to switch the formulations.

SUMMARY

The pharmacologic evaluation conducted by the RTOG in its Phase I studies have indicated essentially 100% bioavailability of misonidazole and no evidence of difference between capsule and tablet form. Serum half-lives were found to be slightly shorter in children than in adults and levels on Fridays slightly higher than on Mondays with daily administration in both groups. Tumor levels were found to be approximately 80-90% of mean plateau serum values. Desmethylmisonidazole was found to be an important metabolite of misonidazole representing 10% of the serum level but 60% of the urine level. There was no evidence for a change in the mean 4-6 hour level caused by phenytoin sodium and dexamethasone but half-life determinations were not made in the presence and absence of these compounds (12). Workman et al. have found a shortening of serum half-life with the compound (13). Extension of the Phase I studies to Phase II and now Phase III has revealed consistence in the serum levels among the large number of member institutions in the RTOG. The few discrepancies have been corrected by inter-comparison studies with a reference laboratory.

Table 1. Misonidazole and Desmethylmisonidazole Analysis: U.V. Spectrophotometric Methods

A. Blood (Serum or Plasma) Determinations

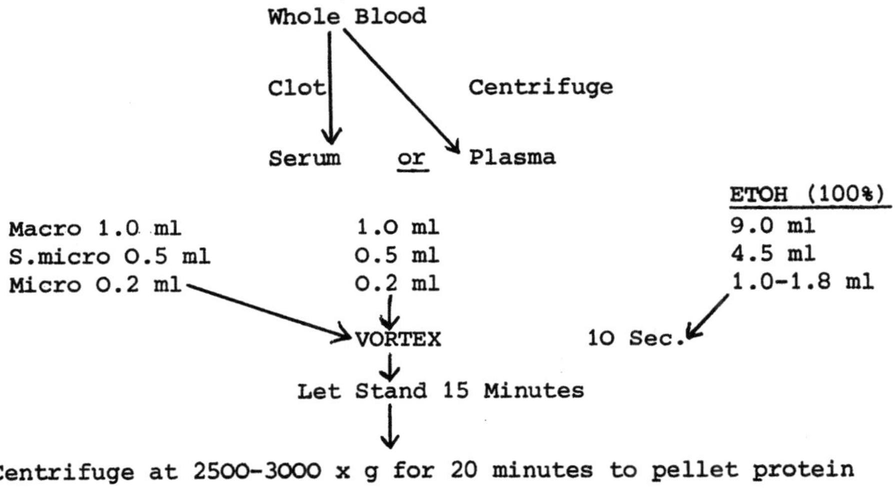

The optical density of the supernatant is measured at 318-325 nm. The exact value should be established for each laboratory at the peak of spectrophotometric absorption of a standard sample. The blank cuvette should contain an identical aliquot of the patient's own serum (plasma) obtained prior to any misonidazole administration and treated in the identical fashion shown above.

Using this method, a standard curve is set up by placing various quantities from 10-200 µg/ml of misonidazole in the blood prior to preparation. The readings of the standard curve are then compared to the reading of the unknown to determine the quantity of misonidazole and desmethylmisonidazole.

B. Precautions and Pitfalls

1. Do NOT use plastic tubes or containers as misonidazole may adsorb to different plastics to a different degree.

2. A variation of up to ± 0.8 µg/ml misonidazole is not uncommon for different reference sera (plasma).

Table 1. Contd.

3. Keep kit samples frozen until used as bacteria grow rapidly and destroy misonidazole. If freshly drawn patients' serum (plasma) cannot for some reason be read at once, it must be refrigerated. If the delay in reading the optical density on the U.V. spectrophotometer is going to be more than a few hours, the sample must be frozen.

4. This method will never distinguish between misonidazole and its primary metabolite, desmethylmisonidazole. Thus, any desmethylmisonidazole will contribute to the misonidazole absorbance. The HPLC assay will separate these compounds.

5. Don't let the blood hemolyze. A pink plasma sample may change the absorbance at 323 nm by 0.1 absorbance units and a red sample will lead to larger errors. The HPLC assay for misonidazole is not sensitive to sample hemolysis.

6. The sample may continue to become turbid or crystals may form after centrifugation. Make sure the sample is clear just prior to reading its absorbance. If not, both the unknown sample and the blank must be recentrifuged again at 2500-3000 x g for 15-20 minutes to remove any additional crystals (uric acid). These microcrystals, which may also form during the storage of the processed serum (plasma) at lowered temperature, will give a false high reading. Cold stored samples must be centrifuged and allowed to return to room temperature before determining the optical density in order to minimize the interference which additional crystallization would impart.

7. Be sure to obtain a sufficient amount of the patient's serum (plasma) prior to the administration of <u>any</u> misonidazole to supply sufficient standards (blanks) to carry out the entire series of UV analyses throughout the complete course of radiotherapy.

8. Remember that both misonidazole and desmethylmisonidazole are quite photosensitive and <u>must</u> be protected from light. Serum (plasma) samples (whether already processed or not) which are being allowed to equilibrate with room temperature must be kept completely shielded from the light.

C. <u>Tissue Determinations</u>

1. Samples of 0.5 to 1 gm of tissue are minced with scissors in 9 (or 4.5 of 1.0-1.8) cc of ethanol (see Flow Sheet in A above). As much blood as possible should be allowed to

Table 1. Contd.

drain from the sample before preparation; then loose blood should be blotted from the surface.

2. **Samples should be allowed to stand for one to two hours.**

3. **Samples are then centrifuged at 2400 x g for 10 minutes.**

4. **The optical density of the supernatant is measured at 318-325 nm. The exact value should be established for each laboratory at the peak of spectrophotometric absorption of a standard sample as outlined above.**

5. **A standard curve is set up in the manner similar to that for blood and used to determine the exact level in the unknown sample.**

Table 2. Misonidazole and Desmethylmisonidazole Analysis (High Pressure Liquid Chromatographic Procedure)

Concentration Range of Analysis:
Misonidazole - 10-200 μg/ml
Desmethylmisonidazole - 5-50 μg/ml

Quantitation of misonidazole and desmethylmisonidazole may be achieved using either an internal standard or external standard method; however, the latter is preferred to maximize accuracy of the measurement.

A. Reference Compounds

1. Misonidazole (1-[2-nitroimidazol-1-yl]-3-methoxy-propran-2-ol)

2. Desmethylmisonidazole (1-[2-nitroimidazol-1-yl]-3-hydroxypropran-2-ol)

3. Internal Standard (1- [2-nitroimidazol-1-yl]-3-chloro-propran-2-ol), (1-[2-nitroimidazol-1-yl]-3-fluoropropran-2-ol), or a 3-alkoxy substituted analog of misonidazole.

B. Instrumental Parameters

1. Column: μ Bondapak C_{18} column (Central Reference Laboratory at NCI uses a Whatman Partisil PXF 5/25 ODS, Whatman, Inc., Clifton, NJ).

2. Eluting Solvent: 5% acetonitrile, 95% 0.45 Macetate buffer, pH 4.5. This solution is prepared by adding 30g anhydrous

Table 2. Contd.

sodium acetate and 30 ml glacial acetic acid to 1500 ml of distilled water to a final volume of 2 litres.

3. Solvent flow rate: 2.0 ml/min.

4. Wavelength: 324 nm.

5. Elution times on NCI Central Reference Laboratory column:

Desmethylmisonidazole	3.0 min
Internal Standard (Fluoro Analog)	5.1 min
Misonidazole	7.7 min

C. <u>Standard Solutions</u>

1. Internal Standard - 10 μg/ml in 100% MeOH.

2. Misonidazole - 10,40,80,120,160,200 μg/ml in pooled human plasma or serum.

3. Desmethylmisonidazole - 5,10,20,30,40,50 μg/ml in pooled human plasma or serum.

D. <u>Blood (Serum or Plasma) Procedure</u>

Whole Blood

clot → serum (0.2 ml)

centrifuge → plasma (0.2 ml)

↓

Add (0.8 ml)

100% MeOH (the extracting solvent contains internal standard if using this method).

↓

VORTEX (15 sec)

↓

Centrifuge at ≃ 2,000 x g for 2 min

↓

Inject 20 μl of supernatant into HPLC for quantitation

E. <u>Calibration Curves</u>

For both the internal and external standard methods of both misonidazole and desmethylmisonidazole, the standard solutions described in item C should be analyzed and calibration curves generated. This is necessary to determine that the analysis is linear over the concentration range expected for the method. Each standard solution should be analyzed as outlined in item D

Table 2. Contd.

and the peak height or area of the recorder response measured. For the external standard method, a calibration curve is constructed by plotting the peak height or area of the reference compound in the sample vs. its known concentration. For the internal standard method, the peak height or area of both the reference compound and that of the internal standard are measured. A calibration curve is then constructed by plotting the peak height or area ratios of either misonidazole or desmethylmisonidazole and that of the internal standard vs. the known concentration of each standard solution. The calibration curves should be linear and pass through the origin. Once linearity has been established, a single standard solution of misonidazole and one of desmethylmisonidazole should be analyzed each day prior to quantitating patients' samples to provide a check on the reliability of the method.

F. Calculations

1. Graphic - After analyzing an unknown sample, the peak heights or areas recorded are measured and the concentration of either misonidazole or desmethylmisonidazole is obtained directly from the respective calibration curves.

2. Calculation from Slope of Calibration Curve - Once the slope of the calibration curve is determined, the concentration of misonidazole or desmethylmisonidazole in the unknown sample can be computed using the following relationships:

For the external standard method:

$$\text{sample concentration } (\mu g/ml) = (\text{slope}) \times (\text{sample peak height or area})$$

For the internal standard method:

$$\text{sample concentration } (\mu g/ml) = (\text{slope}) \times \frac{(\text{sample peak height or area})}{(\text{internal standard peak height or area})}$$

Note: Previous data on patients receiving misonidazole i.v. shows that the ratio of desmethylmisonidazole to misonidazole in plasma averages 0.12 and 0.26, 4 hours and 24 hours, respectively, after dosing with misonidazole.

Table 3. Relationship of Dose to Mean 4-6 hour Serum Level

Dose (g/m^2)	(No. of Doses) (Determinations)	Mean 4-6 hr Serum Level (μg/ml) (0582 + 9963)	+/- Standard Deviation of:	Observed Range
1.0	17	30	+/- 9	21 - 57
1.25	116	46	+/- 6	31 - 61
1.5	78	60	+/- 18	16 - 95
1.75	18	70	+/- 18	39 - 107
2.0	99	76	+/- 14	51 - 120
2.5	43	98	+/- 23	35 - 148
3.0	25	116	+/- 26	39 - 155
4.0	8	175	+/- 56	83 - 238
5.0	7	183	+/- 60	91 - 245

Table 4. Misonidazole Phase I and II Studies: Relationship of Dose to Mean 4-6 Serum Level

Dose g/m^2	No. of Doses (Determinations)		Mean 4-6 hr. Serum Level µg/ml		+/- Standard Deviation		Observed Range	
	Phase I	Phase II	Phase I	Phase II	Phase I	Phase II	Phase I	Phase II
5.0	7	7	183	170	60	47	90-245	80-210
4.0	8	7	175	146	56	44	83-238	68-204
3.0	25	-	116	-	26	-	39-155	-
2.5	43	324	98	83	23	29	35-148	4-165
2.0	99	306	76	68	14	22	51-120	2-141
1.75	18	50	70	64	18	17	39-107	3-102
1.5	78	225	60	57	18	23	16-95	3-179
1.25	116	65	46	44	6	17	31-61	6-74
1.0	17	10	30	49	9	13	21-57	25-72

Table 5. Bioavailability of Misonidazole: Pharmacology - Mean 4-6 hr. Levels

Type of Patient	No. of Patients	Individual Dose g/m^2	Mean 4-6 hr. Level Capsules μg/ml*	Mean 4-6 hr. Level Tablets μg/ml
ADULT°-DAILY-Monday	1	0.4	15	16
Friday	1	0.4	17	18
Monday	7	0.5	19.9 ± 9.6	19.3 ± 4.5
Friday	6	0.5	25.7 ± 11	23.8 ± 7.3
ADULT-Intermittent	2	1.0	52.5 ± 2	45.5 ± 5
	2	1.25	39.5 ± 22	48 ± 20
	5	1.5	59 ± 15	62.4 ± 10
	1	2.0	59	75
CHILDREN+				
Daily Monday	1	0.5	23	18
Friday	1	0.5	25	25
Monday	1	1.0	33	32
Friday	1	1.0	57	51
CHILDREN-Intermittent	3	1.5	61.7 ± 15	65.7 ± 5

°of 34 adults, 25 had comparison blood levels

+of 9 children, 5 had comparison blood levels

* ±standard deviation

Table 6. Bioavailability of Misonidazole: Pharmacology-Half Life

Type of Patient	Individual Dose	$T_{1/2}$ Capsules	$T_{1/2}$ Tablets
ADULT	1 g/m^2		20.5 hr[+]
	1.25 g/m^2	11.5 hr	10.5 hr
	1.5 g/m^2	12.25 hr	14.0 hr
	2.0 g/m^2	8.25 hr	10.25 hr
	2.5 g/m^2	16.0 hr	20.5 hr
mean half life		12 ± 3 hr	13.8 ± 4.8
CHILDREN	1.5 g/m^2	11.25 hr	13.5 hr
		10.5 hr	5.75 hr
mean half life		10.9 ± 5.3	9.6 ± 5.5

[+]First value not used - no comparison

REFERENCES

1. G. E. Adams, I. R. Flockhart, C. E. Smithen, I. J. Stratford, P. Wardman and M. E. Watts, Electron-affinic sensitization. A correlation between structures, one-electron reduction potential, and efficiencies of nitroimidazoles as hypoxic cell radiosensitizers, Radiat. Res., 67:9 (1976).
2. N. J. McNally, J. Denekamp, P. Sheldon, I. R. Flockhart and F. A. Stewart, Radiosensitization by misonidazole (Ro 07-0582). The importance of timing and tumour concentration of sensitizer, Radiat. Res., 73:568 (1978).
3. J. Denekamp, Testing of hypoxic cell radiosensitizers in vivo, Cancer Clin. Trials, 3:139 (1980).
4. S. Dische, A. J. Gray and G. D. Zanelli, Clinical testing of the radiosensitizer Ro 07-0582, II Radiosensitization of normal and hypoxic skin, Clin. Radiol. 27:159 (1976).
5. T. L. Phillips, T. H. Wasserman, R. J. Johnson, The hypoxic cell sensitizer program in the U.S., Br. J. Cancer, 37: 276, Suppl.III (1978).
6. T. L. Phillips and T. H. Wasserman, Hypoxic cell sensitizer studies in the U.S., in "The Treatment of Radioresistant Cancers", Elsevier/North Holland Press, 29 (1979).
7. T. L. Phillips, T. H. Wasserman, R.J. Johnson, Final Report on the United States Phase I clinical trial of the hypoxic cell radiosensitizer, misonidazole. Submitted to Cancer (1980).
8. T. H. Wasserman, T. L. Phillips, R. J. Johnson, Initial U.S. clinical and pharmacologic evaluation of misonidazole, an hypoxic cell radiosensitizer, Int. J. Radiat. Onc. Biol. Phys., 5:775 (1979).
9. T. H. Wasserman, J. Stetz and T. L. Phillips, Clinical Trials of misonidazole in the U.S. In press, Cancer Clin. Trials (1980).
10. R. A. Marques, B. Stafford, N. Flynn and W. Sadee, Determinations of metronidazole and misonidazole and their metabolites in plasma and urine by high performance liquid chromatography, J. Chromatog., 146:163 (1978).
11. G. VanRaalte, T. L. Phillips and T. H. Wasserman, Bio-availability of misonidazole in capsule and tablet form, (abstract) in press, Int. J. Radiat. Onc. Biol. Phys. (1980).
12. T. H. Wasserman, R. Urtasun and J. Schwade, The neurotoxicity of misonidazole: Potential modifying role of phenytoin sodium and dexamethasone, Br. J. Radiol., 53:173 (1980).
13. P. Workman, Effects of pretreatment with phenobarbitone and phenytoin on the pharmacokinetics and toxicity of misonidazole in mice, Br. J. Cancer, 40:335 (1979).

METRONIDAZOLE AND MISONIDAZOLE : ABSENCE OF CYTOGENETIC EFFECTS IN A EUOXIC BACTERIAL AND A MAMMALIAN CELL SYSTEM *IN VITRO*

J.R. Dunlop, J.S. Mahood and R.L. Willson

Biochemistry Department, Brunel University
Uxbridge
Middlesex, U.K.

INTRODUCTION

The drug metronidazole (Flagyl) has been widely used for many years in the treatment of trichomoniasis amoebiasis and giardiasis and is currently proving of considerable benefit in the treatment of anaerobic infections generally (1). Along with a related compound misonidazole (Ro-07-0582) it is also presently undergoing trials as an adjunct in cancer radiotherapy (2-4). Concern about its use has been expressed, however, particularly in the United States because of the reports of its mutagenic and carcinogenic activity in bacterial and animal systems (5-12). The relevance of these reports to the human situation has been the subject of much discussion (1,13). An increase in chromosome aberrations above untreated controls has been found in patients undergoing treatment for Crohn's disease (14) although a similar study in patients being treated for Trichomonas vaginalis no increase was detectable (15). Like metronidazole, misonidazole has also been shown to be active in bacterial and mammalian systems in vitro (16,17).

We now report a further assessment of the cytogenetic activity of metronidazole and some related drugs with particular emphasis on the role of oxygen. Two systems have been used. First a bacterial system similar to the histidine-dependent Salmonella system used in the well-known Ames' test has been employed. Measurements have been made of the ability of metronidazole and misonidazole to cause a relatively stable tryptophan requiring mutant E. coli WP2 to revert to a prototrophic state in which it no longer requires the amino acid for growth. Second, a mammalian cell system has been employed in which the ability of metronidazole, misonidazole and some related compounds to increase the frequency of sister chromatid exchanges has been assessed. Although the genetic significance of this process is not fully understood, it is attracting considerable interest as a possible indicator of the effects of mutagens and carcinogens on eukaryotic chromosomes. It appears that the exchanges, observed by selective staining of the separate chromatids, can be correlated with events such as mutagenesis which are associated with cell survival, in contrast to chromosome aberrations which are often associated with cell death.

EXPERIMENTAL

Metronidazole was kindly supplied by May & Baker Ltd., and Misonidazole by Professor G.E. Adams and Roche Products Ltd.

Bacterial mutation. The treat and plate method used was essentially similar to that described previously (18). Aerobic or anaerobic cultures of *E. coli* WP2 were harvested by centrifugation and resuspended in growth medium to give a final concentration of *ca* 2×10^9 cells per ml. Suspensions were gently bubbled with air or nitrogen for 60 minutes and the drug then added. The number of revertants were obtained by spreading 0.2 ml of undiluted cell suspensions onto triplicate plates containing agar supplemented with tryptophan (0.25 μg/ml) sufficient to allow original mutants to undergo a few divisions but only revertants to form a visible colony. Control samples were also spread onto unsupplemented agar to correct for mutants present in the culture before drug treatment. Additional aliquots of suspensions were sequentially diluted and plated on nutrient agar to determine the number of cells present and the extent of any toxic activity. Mutagenic activity was subsequently calculated in terms of the induced mutation frequency (IMF), the number of revertants above background per 10^8 cells.

Sister chromatid exchange (SCE). Chinese hamster ovary (CHO) cells were cultured as monolayers in glass roller bottles at 37°C in an atmosphere of 5% CO_2. Medium (Ham F.12 Flow Labs. supplemented with 10% foetal calf serum but no antibiotics) was used throughout the study: the cell doubling time was 11 to 15 hours. Asynchronous cells (*ca* 5×10^4) in the exponential growth phase were incubated in the dark on 10 cm dishes for 24 hours in medium containing 1 μM bromodeoxyuridine (BUdR) and various concentrations of drug. In studies of the effect on hypoxic cells (*ca* 5×10^6) were incubated in sealed vials with BUdR and metronidazole for 8 hours the gas space above the medium being previously purged with a 95% N_2 5% CO_2 mixture. Cells were then incubated for 24 hours with BUdR. After incubation, colchicine (0.2 mg/ml) was added to arrest cells in metaphase and samples of mitotic cells removed 2 hours later, by shaking. Sister chromatids were differentially stained using the 'Fluorescence plus Giemsa' technique (19). Air-dried slides were stained for 12 min with Hoechst 33258 (50 μg/ml) rinsed in distilled water, placed in a shallow tray, flooded with

0.1M McIlvaine buffer pH 7.4 and illuminated with a 40W fluorescent lamp for 4 hours. Slides were then rinsed, dried and stained for 6 min with 5% Giemsa made up in M/15 Sørensen buffer (pH = 6.8). Slides were rinsed, air dried, mounted permanently with DPX coded and scored by one observer. A minimum of 20 usually 40 well-spread metaphases were analysed for each data point (20).

RESULTS

Preliminary experiments were undertaken to determine the toxicity of metronidazole and misonidazole towards E.coli WP2. Plots showing the percentage cell survival with time on incubation of cells at different drug concentrations are shown (Figs.1,2). With metronidazole (0.5-1 mM) little toxicity was apparent over 5 hours. However, with 5 mM drug considerable killing was apparent when cells were incubated under nitrogen. In contrast misonidazole

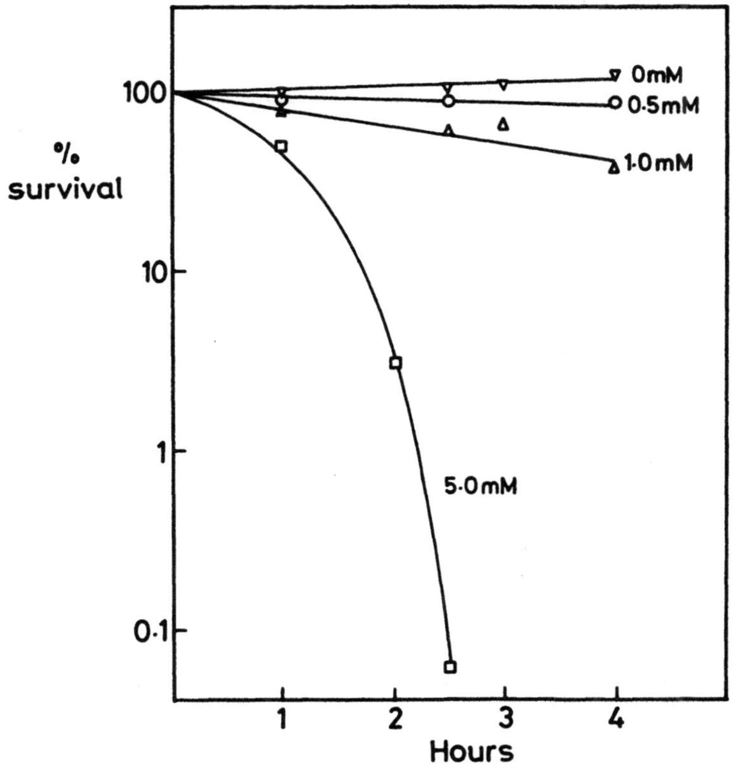

Fig. 1 Viability of E.coli WP2 incubated with metronidazole under hypoxic conditions.

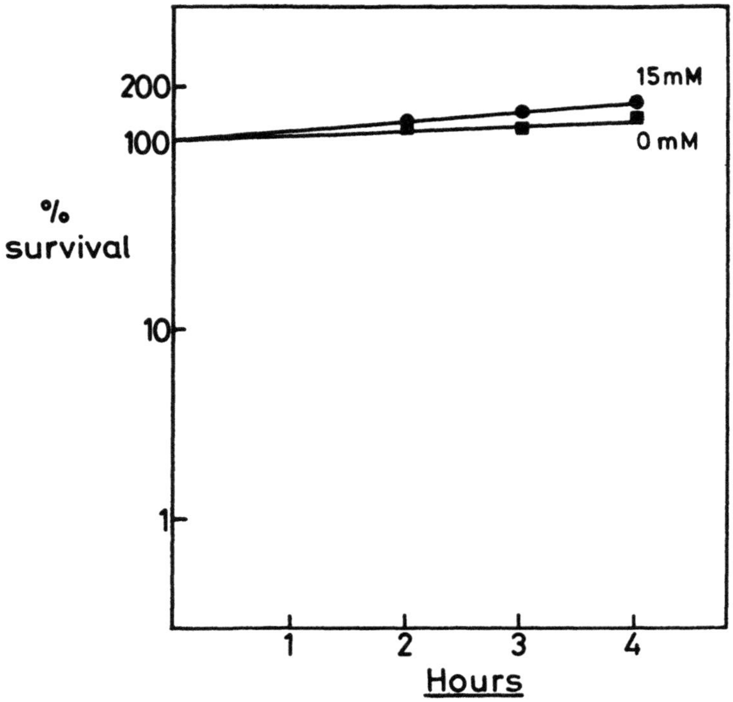

Fig. 2 Viability of E.coli WP2 incubated with misonidazole under hypoxic conditions.

was non-toxic at concentrations up to 15 mM even under nitrogen. Neither drug showed any toxicity when cells were incubated in the presence of air.

In the light of the above toxicity experiments determinations of induced mutation frequency were limited to drug concentrations up to 1.0 mM and 30 minutes after incubation. Again, neither drug showed any activity in the presence of oxygen. Metronidazole, but not misonidazole, showed activity under nitrogen. With metronidazole, the induced mutation frequency increased with increasing drug concentration (Fig. 3).

The frequency of sister chromatid exchanges occurring in CHO cells incubated under air for 24 hours in the absence and presence of metronidazole, misonidazole and other nitroimidazoles in the concentration range 0-1000 μM is shown in table 1. Previous studies had shown that under these conditions negligible cell killing occurred, (plating efficiency > 90%). Examination of the

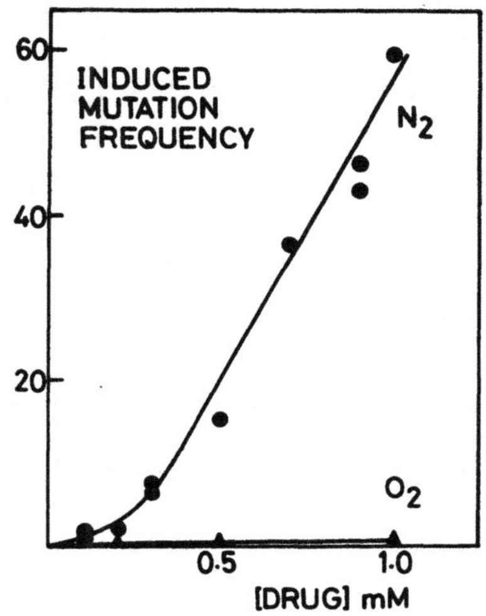

Fig. 3 Induced mutation frequency per 10^8 cells in hypoxic and euoxic E.coli WP2 incubated with metronidazole for 30 mins.

data by analysis of variance using the protocol of Woolf (21) for partitioning sums of squares showed that the difference in frequency between drug treatments and control are not significant ($P > 0.05$). In this system even when cells were incubated with metronidazole under nitrogen for 8 hours no significant increase in frequency was apparent, although the control level was slightly raised. To check that the system could detect an increase in induced mutation frequency confirmatory experiments were undertaken with mitomycin C. When cells were incubated with the drug at a concentration of 0.01μM under air, the induced mutation frequency increased from the background level of 7.0 to 26.4 per cell.

Table 1. Sister Chromatid Exchanges Per Cell

DRUG (24 hrs)	CONTROL	10μM	100μM	1000μM
METRONIDAZOLE	6.8±1.0	7.5±0.8	6.8±1.1	7.3±0.9
MISONIDAZOLE	6.9±1.5	7.0±0.5	7.5±0.5	6.6±0.5
DIMETRIDAZOLE	7.7±0.2	7.7±1.4	6.2±0.4	6.8±0.9
TINIDAZOLE	7.0±1.1	5.5±1.4	5.5±0.4	6.8±0.9
METRONIDAZOLE (N_2, 8hrs)	9.2±0.4	—	—	9.7±0.3
0.01μM MITOMYCIN C (24 hrs): 26.4±1.1				

DISCUSSION

The observed lack of cytogenetic action of any of the drugs studied here under aerobic conditions, is in accord with previously reported studies in which metronidazole failed to induce dominant lethal genes in embryos (22) unscheduled DNA synthesis in cultured human fibroblasts (22) or micronuclei in mammalian cells in vivo (23). The fact that metronidazole in the present study only increased the induced mutation frequency in bacteria incubated under hypoxia is in agreement with previous studies with E.coli and with intestinal bacteria showing that drug reduction only occurs in the absence of oxygen and that nitro group reduction is a necessary prerequisite for damaging activity (24,25). The lack of activity of misonidazole towards E.coli WP2 compared to metronidazole is surprising in view of the fact that misonidazole is reduced more rapidly by these bacteria, but is in agreement with previous studies with Serratia marcescens, E.coli B/r and with protozoa (24,26).

Although the exact mechanism of action of metronidazole remains to be elucidated, interference with electronic transport systems and the formation of toxic products in the absence of oxygen are generally thought to be involved. Iron-sulphur centres or reduced flavins are likely candidates as sites of reduction (27-30). In addition to nucleic acid, toxins produced may damage enzyme systems sensitive to oxidation such as pyruvate dehydrogenase or lactate dehydrogenase nicotinamide adenine dinucleotide complexes, and so interfere with cell metabolism with damaging consequences (29,31). The fact that misonidazole is non-damaging in some bacterial systems suggests that reaction of the toxic products formed with this drug take place at non-critical sites. This would be in agreement with its high electron affinity (39) and the probability of it being less fastidious in its site of reduction (24).

SUMMARY

The cytogenetic activity of metronidazole, misonidazole and some related nitroimidazole drugs have been assessed in bacteria using the mutant E.coli WP2 and in mammalian cells in vitro by measuring the frequency of sister chromatid exchange. Although metronidazole was active in the bacterial system it was so only in the absence of oxygen. Neither metronidazole nor misonidazole showed any cytogenetic activity in the mammalian cell system even when oxygen was absent.

ACKNOWLEDGMENTS

Financial support from the Cancer Research Campaign and the International Atomic Energy Agency is gratefully acknowledged. One of us, JSM, is supported by a SRC. C.A.S.E. Studentship in conjunction with May & Baker Ltd. We are indebted to Dr. M. Green of the M.R.C. Cell Mutation Unit, Sussex for helpful discussion.

REFERENCES

1. E.J. Baines, Metronidazole, Excerpta Medica, S.M. Finegold, ed. p.61 and refs. cited. (1977).
2. C.R. Wiltshire, P. Workman, J.V. Watson and N.M. Bleehen, Clinical studies with misonidazole, Br. J. Cancer 37 Suppl. III, 286 (1978).
3. R.C. Urtasun, J.D. Chapman, P. Band, H. Rabin, C. Fryer and J. Sturmwind, Phase 1 Study of high-dose metronidazole on in vivo and in vitro specific radiosensitizer of hypoxic cells. Radiology 117 :127 (1975).
4. S. Dische and M.I. Saunders, Clinical experience with misonidazole. Br.J. Cancer 37 Suppl.III, 311 (1978).
5. C.E. Voogd, J.J. Van Der Stel and J.J.J.A.A. Jacobs, The mutagenic action of nitroimidazoles, I. Metronidazole, nimorazole, dimetridazole and ronidazole, Mutation Res. 26:483 (1974).
6. M.S. Legator, T.H. Connor and M. Stoeckel, Detection of mutagenic activity of metronidazole and niridazole in body fluids of humans and mice, Science 188:1118 (1975).

7. W.T. Speck, A.B. Stein and H.S. Rosenkranz, Mutagenicity of metronidazole : presence of several active metabolites in human urine, J. Natl. Cancer Inst. 56:283 (1976).

8.a H.S. Rosenkranz and W.T. Speck, Mutagenicity of metronidazole: activation by mammalian microsomes, Biochem.Biophys.Res. Commun. 66:520 (1975).

8.b D.G. Lindmark and M. Muller, Antitrichomonal action, mutagenicity and reduction of nitroimidazoles, Antimicrobiol Agents Chemotherapy 10:

9. J.B. Chin, D.M.K. Sheinin and A.M. Rauth, Screening for the Mutagenicity of Nitro group containing Hypoxic Cell Radio-sensitizers using Salmonella typhimurium strains TA.100 and TA.98. Mutation Res. 58:1 (1978).

10. Tong-Man Ong, B. Slade and F.J. de Serres, Mutagenicity and mutagenic specificity of metronidazole and niridazole in Neurospora. Crassa. J. Toxicol.Environ.Health 4:815 (1978).

11. M. Rustia and P. Shubik, Induction of lung tumors and malignant lymphomas in mice by metronidazole. J. Natl.Cancer Inst. 48:721 (1972).

12. M. Rustia and P. Shubik, Experimental induction of hepatomas, mammary tumors and other tumors with metronidazole in non-inbred Sas:MRC (WI) BR rats. J. Natl. Cancer Inst. 63:863 (1979).

13. C.M. Beard, K.L. Noller, W.M. O'Fallon, L.T. Kurland and M.B. Dockerty, Lack of evidence for cancer due to use of metronidazole. New Eng. J. Med. 301:519 (1979).

14. F. Mitelman, B. Hartley-Asp and B. Ursing, Chromosome aberrations and metronidazole. Lancet October 9th p.802 (1976).

15. B. Hartley-Asp, Mutagenicity of metronidazole. Lancet February 3rd, p.275 (1979).

16. W.D. Rupp, Z. Mroczkowski and K.C. Agrawal, Hypoxic radio-sensitizers: prospects for effective compounds with fewer toxic side effects. Br. J. Cancer 37 Suppl.III, 60 (1978).

17. E.J. Hall, R. Miller, M. Astor and R. Rini, The nitroimidazoles as radiosensitizers and cytotoxic agents. Br. J. Cancer 37 Suppl. III, 120 (1978).

18. M.H.L. Green and W.J. Muriel, Mutagen testing with Trp^+ reversion in Escherichia coli. Mutation Res. 38:3 (1976).

19. P. Perry and S. Wolff, New Giemsa method for the differential staining of sister chromatids. Nature 251:156 (1974).

20. P. Perry and H.J. Evans, Cytological detection of mutagen-carcinogen exposure by sister chromatid exchange. Nature 258:121 (1975).

21. C.M. Woolf, Principles of Biometry. D. Van Nostrand Co. Ltd. pp. 106 and 116.

22. R.G. Bost, Metronidazole. S.M. Finegold ed. Excerpta Medica p.126, (1977).

23. R.J. Trzos, G.L. Petzold, M.N. Brunden and J.R. Stenberg, The evaluation of sixteen carcinogens in the rat using the micro-nucleus test. Mutation Res. 58:79 (1978).
24. S.H. Basaga, J.R. Dunlop, A.J.F. Searle and R.L. Willson, Metronidazole (Flagyl) and misonidazole (Ro-07-0582): reduction by facultative anaerobes and cytotoxic action on hypoxic bacteria and mammalian cells in vivo. Br.J.Cancer 37, Suppl.III, 132. (1978).
25. A.J.F. Searle and R.L. Willson, Metronidazole (Flagyl) degradation by the intestinal flora. Xenobiotica 6:457 (1976).
26. H.N. Prince, E. Grunberg, E. Titsworth and W.F. Delorenzo, Effects of 1-(2-nitro-1-imidazolyl)-3-methoxy-2-propanol and 2-methyl-5-nitroimidazole-1-ethanol against anaerobic and aerobic bacteria and protozoa. Applied Microbiol. 18:728 (1969).
27. R.L. Willson and A.J.F. Searle, Metronidazole (Flagyl): iron catalysed reaction with sulphydryl groups and tumour radio-sensitisation. Nature 255:498 (1975).
28. D. Bahnemann, H. Basaga, J.R. Lunlop, A.J.F. Searle and R.L. Willson, Metronidazole (Flagyl) misonidazole (Ro-07-0582), iron, zinc and sulphur compounds in cancer therapy. Br. J. Cancer 37:Suppl.III 16 (1978).
29. R.L. Willson, Metronidazole (Flagyl) in Cancer Radiology in: "Metronidazole", S.M. Finegold ed. Excerpta Medica p.147 (1977).
30. E.J. Clarke, K.H. Golding and P. Wardman, Anaerobic production of nitroimidazoles by reduced flavin monocleotide and by xanthine oxidase. Biochem. Pharmacol. In press.
31. J.S. Mahood and R.L. Willson, Cytotoxicity of metronidazole (Flagyl) and misonidazole (Ro-07-0582) enhancement by lactate. Br. J. Cancer . Submitted.
32. P. Wardman, The use of nitroaromatic compounds as hypoxic cell radiosensitizers. Curr.Top.Radiat.Res.Quat. 11:347 (1977).

INVESTIGATIONS WITH NITROIMIDAZOLES IN CLINICAL ONCOLOGY

Raul C. Urtasun

Cross Cancer Institute, University of Alberta
Edmonton, Alberta, Canada

ABSTRACT

Clinical investigations on the usefulness of nitroimidazole compounds in cancer therapy were initiated more than six years ago. The first drug to be used was metronidazole. The feasibility of using this compound at doses ten times higher than what was currently used as an anti-trichomonal and anti-amebic agent was assessed by performing Phase I toxicological and pharmacokinetic studies in blood and tumor tissues. At the completion of this initial study it was established that at doses of 6 g/m^2 of surface area, peak plasma levels were obtained in the order of one millimolar with a distribution within tumor tissues of approximately 80-90% of the blood levels. Pilot clinical Phase II trials were initiated in order to assess the efficacy of this agent as a radiosensitizer of hypoxic tumor cells. Shortly thereafter and while these Phase II studies were conducted a new nitroimidazole, misonidazole, was available for toxicity and pharmacokinetic studies. It was soon discovered that misonidazole when given in multiple dose schedules reaching total dose of 11-12 g/m^2, in addition to acute G.I. toxicity, peripheral neuropathies, ototoxicity and CNS symptomatology were the limitations for any further dose escalation. Within that dose limitation this new compound was tested for effectiveness as an enhancer of tumor response when combined with radiotherapy in the hope of improving the therapeutic ratio between normal and tumor tissues. In the interval of four years the studies with misonidazole have gone from initial toxicity and pharmacokinetic studies to Phase II and Phase III clinical studies. Presently there are five multiple institutions, cooperative groups in North America and two in Europe conducting clinical trials in order to assess its value as a radiosensitizer. This, in addition to at least 15 large medical centers in both North

America and Europe that are conducting specific studies. Close to 1,000 cancer patients have already been treated with metronidazole or misonidazole used in high doses. Early results have shown an encouraging trend and it is expected that within the next one to two years definitive results on the Phase II and III clinical trials will become available. In the meantime, new drugs with the same or more effectiveness and likely less toxic are becoming available for initial clinical investigations. One of these drugs, the metabolite of misonidazole, desmethyl misonidazole, is presently available for clinical toxicity and pharmacokinetic studies.

In addition to their use as radiosensitizers of hypoxic tumor cells the nitroimidazoles are being tested as enhancers of cancer chemotherapeutic agents in man. Their pharmacological interaction and possible improvement of the therapeutic ratio in tumors when used with agents such as Cyclophosphamide and 5-FU are currently being assessed.

INTRODUCTION

Chronically hypoxic cells are present in solid tumors of animal tumor models and in man (1-4). Whether or not the radioresistance of such cells is a limiting factor in the local control of these solid tumors is the subject of current debate. If it is, different approaches are potentially available to lead to improved results. One of them, the use of chemical sensitizers, appears to be promising because of its simplicity, low cost and easy availability.

For the past several years, considerable interest has been directed towards the nitroimidazole class of drugs which has been found to have chemotherapeutic and radiosensitizing properties against hypoxic cells <u>in vitro</u> (5-10) and acceptable pharmacological properties in animal models (11). Two of these nitroimidazole compounds (misonidazole and metronidazole) have been investigated in Phase I and II clinical trials and are undergoing Phase III trials in both Europe and North America. The use of these nitroimidazole compounds, either in combination with conventional radiation or large fractions of radiation, in combination with local hyperthermia and with presently available cancer chemotherapeutic drugs is undergoing extensive clinical evaluation.

METHODOLOGY

Available Formulations

Both oral and parenteral formulations are available for metronidazole and misonidazole (Table 1). The pharmacokinetic studies have been performed in blood and tumor tissues for both the oral (12-15) and parenteral administration (16-17). The most current available methodologies for the clinical measurement of the drug in

Table 1. Available Formulations for Currently used Nitroimidazole Compounds

FORMULATION FOR CLINICAL INVESTIGATIONS		
Metronidazole		
Oral	:	500 mg tablets
I.V.	:	500 mg/100 ml solution
Misonidazole		
Oral	:	500 mg and 100 mg tablets and capsules
I.V.	:	500 mg/20 ml solution
Desmethyl Misonidazole (9963)		
I.V.	:	500 mg/10 ml solution

blood, tissues and cerebral spinal fluid are the spectrophotometric assay, the polarographic assay and the high performance liquid chromatographic assay (HPLC). The HPLC method is preferable since it differentiates between the primary compound and its metabolite. The spectrophotometric assay is fast and could be used in urgent situations where levels of drug are needed immediately and no HPLC is available.

The areas of interest where work has been focussed are: a) blood kinetics and drug metabolism; b) agents that affect the drug metabolism; c) tumor-tissue kinetics in relation to kinetics of the drug in blood; d) timing of radiation treatment in relation to the blood and tumor tissue drug levels; e) toxicity of these compounds and its possible modification by drugs that affect its metabolism; f) interaction of mitroimidazoles with cancer chemotherapeutic agents in regard to the kinetics of either compound; g) studies on drug efficacy such as Phase II pilot trials aiming to demonstrate a sensitizer enhancement ratio or Phase III clinical trials to compare the results of this new modality with the results of conventional well-established methods of treatment.

RESULTS

Extensive pharmacokinetic studies have been performed with the oral and parenteral administration of high dose metronidazole and misonidazole. The results have demonstrated that with the oral formulation of either of these compounds peak plasma levels occur two to four hours after administration with the plasma half-life ranging between 9 and 17 hours. Tumor tissue kinetic studies have established that a great variability occurs from tumor to tumor, although in general it can be said that drug concentrations of anywhere between 60 and 90% of the blood values can be easily obtained in tumor tissues. In our experience the drug appears to distribute uniformly in all compartments of solid tumors. Although it has been suggested by other studies (18) that reasonably high levels of the drug cannot be obtained in the necrotic areas of the tumor. Both in our studies and in other centers the measurements were done of the bound and unbound drug. Studies are currently being conducted in our Institution to measure only the bound drug and so far these findings are suggestive that the cell bound drug distributes only in the hypoxic but viable intermediate zone with no evidence of concentration in either the necrotic areas or in the well-oxygenated regions.

The time of maximum tumor tissue levels does not correlate with the time of maximum blood levels showing generally a lag-time of 2-6 hours. The disappearance of the drug in tumor tissues parallels the disappearance of the drug in blood. Tissue levels of one of the metabolites of misonidazole, desmethyl misonidazole were found in tumor tissue in lower proportions than in blood. Only the ethoxic metabolite of metronidazole was found to be present in tumor tissues and again at lower levels than in blood. Recognizing this lag-time it has been concluded that a period of time between 2 to 4 hours longer than the time of maximum blood concentrations should be allowed before proceeding with radiation treatment.

When either metronidazole or misonidazole are administered intravenously, the alpha $t_{1/2}$ has been calculated to be 8 minutes and that for beta phase $t_{1/2}$ has been calculated to be 6 hours (16-17). Time of maximum tumor tissue concentration of the drug when administered intravenously depends on the speed of the administration as well as its volume. We have found that with the constant infusion time interval of 30 minutes there was still a 3-to-4 hour lag-time between blood and tumor tissue similarly as was seen with the oral preparation (16). However, more extensive studies have been done with misonidazole (17), in which the medication was administered over a shorter period of time and in a smaller volume. Higher concentrations in both blood and tumor tissues were found with almost no lag-time between maximum levels in blood and tumor tissues.

Attempts have been made also to administer these compounds into natural cavities such as the intravesical administration of misonidazole (19) and intrarectally (J. Pedersen, personal communication). Levels of the drug in carcinoma of the bladder, drainage nodes, normal tissues and lymph nodes were determined. The findings suggest that there is a higher concentration of the drug in both lymph nodes and tumor tissues with a much lower concentration in normal tissues. This route of administration will certainly offer its advantages with certain types of tumors.

Acute toxicity with the single oral formulation is mainly G.I. tract in the form of nausea and vomiting. They all occur hours after drug administration and are transient. Multiple dose schedules revealed that the dose-limiting toxicity is either acute G.I. toxicity, sensory peripheral neuropathy, ototoxicity and CNS symptoms of encephalopathy (ataxia, confusion, disorientation, tremors and occasionally convulsions). The peripheral neuropathy observed with both misonidazole and metronidazole are of a mixed type (sensory and motor) with the major component being sensory. It is known that in man, the pathological findings are of peripheral nerve demyelination and axonal degeneration (20). Since a decrease in the incidence of peripheral neuropathies was seen in patients receiving concomitant dexamethasone and phenytoin, pharmacokinetic studies were initiated in several centers to assess the potential use of these drugs to reduce the toxicity of nitroimidazoles either by modifying the kinetics of the drug or by modifying the already established cell membrane damage (21).

Although pilot clinical experimental work has shown encouraging results suggestive of an enhancement of radiation response in artificially hypoxic normal tissues (22) in small subcutaneous tumor nodules (23-24) as well as in one early clinical Phase II study in brain tumors (25) there has not yet been more conclusive results from the on-going Phase II and III clinical studies.

DISCUSSION

Up to the present time the mechanism of the peripheral and central nervous tissue damage is not known although some theories have been postulated such as the inhibition of neuronal protein synthesis secondary to binding of metabolic products into RNA (26-27).

It is apparent that by administering the medication parenterally the acute G.I. toxicity as well as peripheral neuropathies has not been circumvented. One can conclude also, that plasma levels following high dose intravenous administration of nitroimidazoles are less influenced by interfering absorption patterns and therefore a truer distribution and elimination curve results in more accurate kinetic estimates. With the advent of new nitroimidazoles which are more hydrophilic than those presently available, more experience will be needed in the future in the intravenous use of these drugs.

It could be assumed that some of the reasons that more definitive evidence of the effectiveness of these drugs as radio-sensitizers has not yet been reported is that most of these trials have been started not more than two years ago and more time is needed for patient accrual and analysis of their survival. It is also likely that the reason that some of the recent preliminary results are not showing a big therapeutic effect in the sensitizer group is that either the fractionation of radiation or the administration of the drug is not the optimal one. Further studies are needed to establish the optimal scheduling of administration of the radiation and the drug. As well, there is a need for further studies in order to investigate the feasibility of decreasing the toxicity of these drugs, so higher doses can be administered with the radiation. In the meantime, new drugs with equal or more effectiveness and less toxicity are becoming available for clinical testing. One of them, the desmethylated metabolite of misonidazole (RO-05-9963) has shown equal sensitizing effectiveness in vitro and in vivo but less toxicity and is now undergoing early clinical studies.

Recently, great interest has been developed in using nitroimidazoles as enhancers of cancer chemotherapy agents when used a few hours prior or simultaneously with agents such as Cyclophosphamide, Adriamycin, 5-FU and BCNU. The mechanism of this interaction is not known at the present time and studies are being conducted to assess the possible changes in pharmacokinetics of either misonidazole or the cancer chemotherapeutic agent when used together. From the animal studies done to date, the enhancement effect is seen mainly in the tumor with only a small amount of enhancement seen in normal tissues. This of course depends strictly not only on the type of experimental tumor but in the use of a chemotherapeutic agent which is already proven to be effective against that particular tumor (Minutes, Radiosensitizers/Radioprotectors Group, N.C.I. (U.S.A.), Bethesda, July 1980).

ACKNOWLEDGMENT

This work was supported by Grant #55-45033 of the National Cancer Institute of Canada and Alberta Heritage Fund for Medical Research.

REFERENCES

1. J.F. Fowler, G.E. Adams and J. Denekamp, Radiosensitizers of hypoxic cells in solid tumors. Cancer Treatment Reviews 3:277 (1976).
2. R.P. Hill and R.S. Bush, A lung colony assay to determine the radiosensitivity of the cells of a solid tumor. Int.J. Radiat. Biol. 15:435 (1969).
3. R.P. Hill, R.S. Bush and P. Yeung, The effect of anaemia on the fraction of hypoxic cells in an experimental tumor. Br.J. Radiol. 44:299 (1971).
4. A.M. Rauth and K. Kaufman, In vivo testing of hypoxic radiosensitizers using the KHT murine tumor assayed by the lung-colony technique. Br.J. Radiol. 48:209 (1975).
5. J.D. Chapman, A.P. Reuvers and J. Borsa, Effectiveness of nitrofuran derivatives in sensitizing hypoxic mammalian cells to x-rays. Br.J.Radiol. 46:623 (1973).

6. J.L. Foster and R.L. Willson, Radiosensitization of anoxic cells by metronidazole. Br.J.Radiol. 46:234 (1973).
7. A.M. Rauth, In vivo testing of hypoxic cell radiosensitizers. in Radiation Research. Biomedical, Chemical and Physical Perspective. Proc. 5th Internat.Congr. Radiation Res. O.I. Nygaard, H.I. Adler and W.K. Sinclair, eds., Academic Press, New York (1976).
8. A.P. Reuvers, J.D. Chapman and J. Borsa, Potential use of nitrofurans in radiotherapy. Nature 237:403 (1972).
9. R. Sutherland, Selective chemotherapy of non-cycling cells in in vitro tumour models. Cancer Res. 34:3503 (1974).
10. R. Sridhar, C. Koch and R. Sutherland, Cytotoxicity of two nitroimidazole sensitizers in in vitro tumor models. Int.J.Radiat. Oncology, Biol.,Phys. 1:1149 (1976).
11. J.C. Asquith, J.L. Foster, R.L. Willson, R. Ings and J.A. McFadzean, Metronidazole (Flagyl). A radiosensitizer of hypoxic cells. Br.J.Radiol. 47:474 (1974).
12. A.J. Gray, S. Dische, G.E. Adams, I.R. Flockhart and J.L. Foster, Clinical testing of the radiosensitizer RO-07-0582. Dose tolerance, serum and tumor concentration. Clin.Radiol. 27:151 (1976).
13. R.C. Urtasun, J.D. Chapman, P.R. Band, H.R. Rabin, C. Fryer and J. Sturmwind, Phase I study of high dose metronidazole. An in vivo and in vitro specific radiosensitizer of hypoxic cells. Radiology 3:129 (1975).
14. R.C. Urtasun, P.R. Band and J.D. Chapman, Phase I study of the hypoxic cell radiosensitizer RO-07-0582. Radiology 122:801 (1977).
15. T.H. Wasserman, T.L. Phillips, R.J. Johnston, C.J. Gomer, G.A. Lawrence, W. Sadee, R.A. Marques, V.A. Levin and G. VanRaalte, Initial U.S. clinical and pharmacological evaluation of misonidazole, a hypoxic cell radiosensitizer. Int.J.Radiat.Oncology Biol. Phys. 5:775 (1979).
16. H.R. Rabin, R.C. Urtasun, J. Partington, D. Koziol, M. Sharon and K. Walker, High dose metronidazole: pharmacokinetic and bioavailability using an intravenous preparation. Application of its use as a radiosensitizer. Cancer Treatment Reports (in press).
17. J.G. Schwade et al, Intravenous misonidazole (NSC 26 1037). Proc. 21st Ann.Scient.Meet. of American Association of Ther. Radiol., New Orleans,(October 1979).
18. T.A. Ridge, S. Dische, M.I. Saunders, M. Stratford and A. Minchinton, The influence of necrosis on the concentration of misonidazole in human tumors. Conf. on Combined Modality Cancer Treatment : Rad. Sens. & Protect. Key Biscayne, Florida, October 1979. (in press)
19. H.K. Awwad, M.M. El-Merzabani and M.V. Burgers, Penetration of misonidazole after intravesical administration in cases of carcinoma of the bilharzial bladder. Br.J. Cancer 37: Suppl.III, 297 (1978).

20. R.C. Urtasun, J.D. Chapman, M.L. Feldstein, P.R. Band, H.R.Rabin, A.F. Wilson, B. Marynowski, E. Starreveld and T. Shnitka, Peripheral neuropathy related to misonidazole: incidence and pathology. Br.J.Cancer 37: Suppl.III, 271 (1978).
21. T.H. Wasserman, T.L. Phillips, G. VanRaalte, R.C. Urtasun, J. Partington, D. Koziol, J. Schwade, D. Gangji and J.M.Strong, The neurotoxicity of misonidazole: potential modifying role of phenytoin sodium and dexamethasone. Br.J.Radiol. 53:172 (1980).
22. S. Dische, A.J. Gray and G.D. Zanelli, Clinical testing of the radiosensitizer RO-07-0582. II. Radiosensitization of normal and hypoxic skin. Clin. Radiol. 27:159 (1976).
23. D. Ash, M.J. Peckham and G.G. Steel, Quantification of tumor response. Growth delay studies in patients with measurable metastases. Proc. 9th L.H. Gray Memorial Conf. Cambridge, England. Br.J.Cancer p.l 4 Sept. 1979 (Abstract) (in press).
24. R.C. Urtasun, P.R. Band and H. Ferri, Tumor growth delay studies in patients with multiple metastatic nodules: Practical difficulties. Int.J.Radiat.Oncology Biol.Phys. 6:875 (1980).
25. R.C. Urtasun, P.R. Band, J.D. Chapman, M.L. Feldstein, B. Mielke and C. Fryer, Radiation and high dose metronidazole in supratentorial glioblastomas. New.Eng.J.Med. 294: 25, 1364 (1976).
26. W.G. Bradley, I.J. Karlsson, G.C. Rassol, Metronidazole neuropathy. Brit. Med. J. 2:610 (1977).
27. R.M. Ings, J.A. McFadzean and W.E. Ormerod, The mode of action of metronidazole in trichomonas vaginalis and other micro organisms. Biochem. Pharmacol. 23:1421 (1974).

PHARMACOKINETIC AND THERAPEUTIC STUDIES WITH MISONIDAZOLE AND DESMETHYLMISONIDAZOLE IN MAN

Norman M. Bleehen

University Department and Medical Research Council Unit
of Clinical Oncology and Radiotherapeutics
Cambridge CB2 2QH
England

INTRODUCTION

The role of hypoxic cell radiosensitisers in the management of cancer in patients has been the subject of several meetings (1,2) including the present one. This report concerns itself with some clinical studies carried out by the author and his colleagues at Cambridge, England, principally with misonidazole (Ro 07-0582 Roche Laboratories, N.S.C. 261037) and more recently with its O-demethylated metabolite, desmethylmisonidazole (Ro 05-9963, Roche Laboratories, N.S.C. 261036).

Studies on the plasma pharmacokinetics of misonidazole and the influence on it of hepatic microsomal enzyme inducers such as phenytoin and phenobarbitone are presented. The concentration of misonidazole achieved in tumours and in particular those of the brain have been studies as part of a series of randomised clinical trials on the use of misonidazole. The results of one small study in grade 3 and 4 astrocytomas which has now been completed, are discussed. Some results of these investigations have already been reported elsewhere (3-6).

Desmethylmisonidazole is also an active radiosensitiser in model systems (7-9), and has recently become available for limited clinical trial. We have commenced Phase 1 studies with the drug and our preliminary data are also presented.

METHODS AND MATERIALS

Nitroimidazole Assay

Reverse-phase high performance liquid chromatography (10) was used for all estimations of the concentration of misonidazole or desmethylmisonidazole in fluid and tissue samples. Heparinised blood samples were centrifuged at $4^{o}C$ to obtain plasma. Tissue samples were kept on ice immediately after biopsy until homogenised in double distilled water by ultrasonic disintegration using an MSE 150 watt ultrasonic disintegrator. All samples were stored in the dark at $-20^{o}C$ until analysed. Full experimental details have been reported elsewhere (3,4).

Pharmacokinetic Studies

All patients gave informed consent when agreeing to participate in these investigations. For a complete study various blood samples were taken at 0.5, 1, 2, 3, 4, 5, 6, 12 and 24h., but this was not feasible in all patients. When sufficient values were available, it was possible to calculate both the half life ($t_{\frac{1}{2}}$) and the area under the curve ($AUC_{0-\infty}$). However, for many patients only values at around the time of treatment or tumour biopsy could be obtained. Doses of misonidazole ranged between 0.45 and $3g/m^2$ with a total dose never exceeding $12g/m^2$ over a minimum of three weeks. Desmethylmisonidazole was given as single doses of $1g/m^2$. Both drugs were given by mouth after a light breakfast. Results are expressed as ±1 standard deviation and statistical analyses for differences between plasma nitroimidazole levels was by student test.

RESULTS

Misonidazole Pharmacokinetics

There is a clear correlation between the administered dose of misonidazole, as g/m^2, and the peak and 4h plasma concentrations of the drug in 128 patient investigations (correlation coefficients of 0.774 and 0.873 respectively). These mean values are presented in Table 1 for patients who are not known to have been taking microsomal enzyme inducing drugs such as phenytoin or phenobarbitone. The relationship is almost as clearly defined when the dose is given as mg/kg body weight (6). We have not seen any significant difference in the concentrations achieved after repeated dosage by the drug either given as a tablet or capsule preparation (3).

The mean $t_{\frac{1}{2}}$ of plasma misonidazole in patients not on anticonvulsants was 11.5 ± 3.8h. This does not appear to be dose dependent over the range of doses used (0.4 - 3 g/m^2). However, in a series of 16 patients with malignant gliomas, who were being

Table 1. Relationship between plasma misonidazole concentrations and administered oral dose at peak and 4h times. Patients were not on known enzyme inducers.

Misonidazole dose g/m^2	Plasma Misonidazole Concentration (μg/ml)	
	Peak Time	4h
0.5	20.6 ± 4.1	15.4 ± 2.6
0.6	30.8 ± 9.3	24.4 ± 5.8
1.5	70.3 ± 14.4	62.3 ± 15.3
3.0	116.3 ± 47.5	101.9 ± 34.3

given dexamethasone and either or both anticonvulsants, phenobarbitone or phenytoin, we observed that the $t_{½}$ was reduced to 8.6 ± 0.6h (6). There was no difference between the peak plasma misonidazole concentrations achieved for a comparable dose ($3g/m^2$) in either group. The 4h misonidazole concentrations for 23 patients on dexamethasone but not on anticonvulsants, was 103.1 ± 31.8 μg/ml and for 15 patients on anticonvulsants as well, was reduced to 87.1 ± 20.3 μg/ml. However, this difference is not statistically significant ($P < 0.09$). There was a significant increase in the plasma concentration of desmethylmisonidazole from 8.0 ± 2.6 in 21 patients to 13.9 ± 5.4 μg/ml in 14 patients ($P < 0.001$) suggesting increased metabolism of misonidazole in the patients on anticonvulsants.

Both phenobarbitone and phenytoin are known to induce hepatic microsomal enzymes which are involved in the O-demethylation of misonidazole to desmethylmisonidazole. This is one of the main pathways for metabolism of the drug and about half of the misonidazole given by mouth can be recovered in the urine as desmethylmisonidazole. Animal experiments with mice in our laboratory (11) clearly showed that enzyme induction by phenobarbitone or phenytoin reduces the $t_{½}$ of misonidazole without affecting peak tumour concentrations of the drug.

We confirmed the relevance of these observations by a study on 6 volunteer patients in whom we determined the plasma pharmacokinetics of misonidazole before and after a 14 day course of phenytoin 100 mg t.d.s. Full details have been published elsewhere (5). In brief, the results showed that there were no significant changes in the peak and 4h plasma misonidazole concentrations. By contrast, there were significant decreases in the mean $t_{\frac{1}{2}}$ from 11.1 ± 3.7 to 7.5 ± 1.4h ($P < 0.001$) and the $AUC_{0-\infty}$ from 2030 ± 643 to 1367 ± 261 μg/ml x h ($P < 0.01$). The effect on O-demethylation was confirmed by significant rises in the desmethylmisonidazole concentrations at peak and 4h times.

We have recently completed a similar series of experiments in which 7 volunteer patients were given phenobarbitone 30mg 3 times a day. There was around a 20% reduction in the plasma misonidazole $t_{\frac{1}{2}}$ from the control pre-induction concentrations. This reduction was maximal by one week after starting to take the phenobarbitone and was not increased after a further 2 weeks. It is of interest to compare the mean reduction after 14 days of 21.4 ± 4.7% with that after a similar time on phenytoin in the previously reported series (5) of 30.7 ± 3.7%.

The significance of these observations for clinical therapy are still not clear. We were alerted to the possibility of the effect by reason of the fact that we had not observed any significant neuropathic complications associated with misonidazole therapy at a total dose of 12 g/m^2 in our glioma study patients. On the basis of the reported experience (12),and our own in patients not on anticonvulsants, we would have expected to have seen several cases. It seems reasonable to explain this effect by the reduction in the AUC of plasma misonidazole caused by the anticonvulsant therapy received by the glioma patients.

An additional explanation may relate to the fact that all patients were receiving dexamethasone. This has also been reported to reduce the incidence of misonidazole neurotoxicity but not the plasma misonidazole $t_{\frac{1}{2}}$ (13). Experiments in mice have confirmed the lack of a significant effect of dexamethasone on plasma $t_{\frac{1}{2}}$ but have shown a small reduction in the drug concentration in normal brain tissue (14).

One may speculate as to the feasibility of deliberately attempting to reduce the incidence of neurotoxicity in patients receiving misonidazole by using such drugs. However, the administration of either steroids such as dexamethasone or anticonvulsants such as phenytoin or phenobarbitone, are not without risk. One cannot envisage the routine use of such drugs in all patients. However, if newer hypoxic cell radiosensitisers do not reach clinical practice without major toxicities, then it may be desirable to try increasing misonidazole dosage by the use of such strategems. It is also possible that there may be other less toxic but equally efficient hepatic microsomal enzyme inducers which could be employed.

The determination of plasma pharmacokinetics requires repeated blood sampling. We therefore investigated the possibility of using saliva samples to measure the plasma $t_{\frac{1}{2}}$ (4). In a series of 10 patients, at misonidazole doses of 0.5 - 3.0 g/m^2, a dose correlation was found between the salivary and plasma misonidazole concentrations. However, there are still some inter-patient variations and we have not used this technique as a routine method, although it could be of value in situations where repeated blood sampling may present problems.

We have had the opportunity to study 2 patients who have had one kidney removed some time before and a third who had one transplanted kidney. The $t_{\frac{1}{2}}$ in these patients were not significantly outside our normal range although two of them were on the high side (Table 2). Plasma urea and creatinine concentrations were at the upper limit of normal in two of these patients and within normal limits in the third. It is of interest that significant neuropathy developed in two of these patients. One had abnormal renal function but a $t_{\frac{1}{2}}$ well within normal limits and the other a normal renal function but a $t_{\frac{1}{2}}$ towards our upper limits of normal.

Peripheral neuropathy has not been a major problem in our practice probably due to the selective nature of patients treated and limitation to a total dose of 12 g/m^2. Most patients have either been receiving dexamethasone or anticonvulsants. Apart from minor degrees of neuropathy and those observed in the above patients with one kidney, our most significant effects have been in the only two patients with head and neck cancer we treated with misonidazole.

Table 2. Plasma $t_{\frac{1}{2}}$ misonidazole in three patients with only one kidney

Patient	Drug dose g/m^2	Total dose g/m^2	$T_{\frac{1}{2}}$ (h)
A.M.[1]	1.5	10.5	10.9
T.N.[1,2]	2.0	10.0	14.6
F.F.	1.5	1.5	15.2

[1] Developed neuropathy
[2] Transplanted kidney

Tissue Misonidazole Concentration

Studies on biopsy specimens of tumour taken at times after misonidazole have suggested that an optimum useful plasma concentration in tumour is reached around 4h after oral administration of the drug (15,16). Our limited studies on 3 patients (one each with primary adenocarcinoma of breast, secondary adenocarcinoma of colon and undifferentiated lung cancer confirmed a similar situation (3). Most clinical studies therefore now involve radiation treatment around 4 - 5h after oral misonidazole.

There is much less information available about misonidazole concentration in brain tumours. We have recently completed a study on 31 patients from whom biopsies were obtained during surgery after misonidazole administration. Because repeated samples cannot be obtained in the same patient, it is not as easy to present data in terms of the optimum treatment time for an individual patient. It is, however, possible to compare the tissue/plasma misonidazole concentration ratios in different patients with biopsies taken over a wide range of times after the drug. On occasion it has also been possible to obtain samples from normal brain, cyst fluid or cerebrospinal fluid (CSF). Detailed results will be reported elsewhere but a brief summary is given in Table 3 and illustrated for patient A in the top panel of Fig.1. A wide range of results is seen but on average highest tumour/plasma ratios were seen between 3 - 5h. Peak plasma concentrations were seen at 1.5 - 2h although the difference at 4h was not large. Consequently, in terms of optimal tumour misonidazole concentrations, radiation treatment at around 3 - 5h after oral administration of the drug seems appropriate.

Table 3. Ratio of tissue/plasma misonidazole concentration in brain tumour patients at various times after administration of misonidazole. Values are expressed as a mean percentage for (n) patients.

Tissue	Ratio of tissue/plasma misonidazole concentration (%) at a median biopsy time in hours				
	2h	3h	4h	5h	6h
Tumour	65 (3)	68 (10)	76 (7)	80 (4)	44 (3)
Brain	69 (3)	81 (8)	52 (2)	91 (3)	85 (1)
Cyst Fluid	-	34 (3)	31 (1)	-	-
C.S.F.	-	99 (3)	70 (1)	-	104 (1)

It is of interest to note that the mean concentrations in normal brain were similar to those seen in tumour although this relationship varies considerably from patient to patient. Concentrations in CSF also varied but with one exception were virtually the same as that in plasma. Of interest was the relatively low concentration seen in the cyst fluid taken from tumours in 4 patients.

These findings suggest that it is necessary to make direct estimations of misonidazole concentrations or indeed of other radio-sensitisers in the tumour tissue of interest rather than assume that CSF or cyst fluid can give a reasonable approximation to that concentration. The arbitrary selection of a treatment time of 4h after misonidazole in our glioma study seems to have been fortunate and probably not too far off the peak tumour concentration, although we have not been able to assess this time with precision.

CAMBRIDGE GLIOMA STUDY

Misonidazole is undergoing study in a number of clinical trials around the world. Interest in its use in malignant brain tumours has been engendered by the extremely poor prognosis of patients with the higher grade tumours receiving currently accepted therapy by surgery and radiation. A report showing some small transient improvement in the results of treating patients with glioblastoma multiforme using radiotherapy and metronidazole (17) led to optimism that misonidazole might be as, or even more, effective.

The Cambridge glioma study was commenced in 1976 and closed to entry of new patients at the end of 1978. Details of this protocol have been reported previously (3,6). 55 patients over the age of 18, with histologically proven supra-tentorial grade 3 and 4 astrocytomas, were randomised to receive one of three treatment schedules. Two control radiation schedules were given without radiosensitiser and the one with misonidazole as follows :

Group 1 : 5656 rad in 28 fractions of 202 rads/fraction over 5½ weeks. 120 patients were allocated to this treatment.

Group 2 : 4352 rad in 12 fractions over 4 weeks, given so that the largest fraction was on Fridays as follows : Monday 294 rad; Wednesday 294 rad; Friday 500 rad. There were 18 patients treated on this schedule.

Group 3 : The same schedule as for Group 2 but in addition misonidazole was given 4h before the 500 rad fractions on Fridays at an oral dose of $3g/m^2$. The total cumulative drug dose was thus $12g/m^2$. 17 patients were randomised to this group.

The radiation doses in both schedules 1 and 2 were calculated to be the equivalent of 1702 ret.

There were no major differences in the distribution of patients between the groups with respect to age, sex, initial performance status, site of disease, extent of surgery and pathological grade. All patients received dexamethasone 2mg three times a day during and for a period after treatment.

The results are disappointing in that all patients have now died. There is no obvious difference in the survival curves of the three treatment groups. The median survivals are : Group 1, 251 d; Group 2, 220 d; Group 3, 270 d.

The failure to demonstrate any improvement in our study is perhaps surprising in view of the favourable report already cited in the studies using metronidazole with a smaller number of patients (17). This latter study, in spite of the reservations about the low radiation dose, would tend to confirm the likelihood of significant radiobiological hypoxia in gliomas. The usual presence of extensive areas of necrosis also suggest the validity of this assumption.

One must look therefore for other reasons to explain our results. Firstly, if the potential gain is not large then the numbers of patients may well have been far too small. Thus, in the current Medical Research Council (MRC) Glioma Study, the planned entry is 400 patients into two treatment groups. Secondly it is by no means certain that the two radiation dose schedules were biologically the same as far as the central nervous system or tumour is concerned, as we do not have suitable data available. The use of the NSD concept based on equivalence of doses for subcutaneous tissue and skin may be inappropriate for the nervous system. There are, in fact, some data to suggest that this discrepancy may well exist (18,19).

Our study was based on the concept that maximal radiosensitisation would be achieved with large doses of misonidazole and hence high drug concentrations in tumour at the time of treatment. This necessitated treatment with misonidazole on only a few occasions in order not to exceed a tolerance dose of the drug. This resulted in the unorthodox fractionation schedule selected. It has now been claimed that multiple small doses of the sensitiser with a larger number of radiation fractions will achieve the same enhancement effect (20). Several new studies, including the current MRC trial, have opted for this type of schedule.

It seems unlikely the results can be explained on the basis of an inadequate concentration of sensitiser in tumour as our tumour biopsy study confirmed reasonable drug penetration at around the time selected for treatment. It is still possible that a slightly higher tumour concentration would have been employed if the treatment time was 3 - 4h rather than 4 - 5h but our data are not that precise.

We have also demonstrated similar penetration of drug into normal brain tissue. Although misonidazole is not thought to sensitise euoxic tissues (21), it remains a possibility that radiation damage could be enhanced in normal nervous tissue. This has indeed been reported for the spinal cord of anaesthetised rats (22) but in this situation may relate to anaesthetic induced hypoxia.

There is, therefore, no clear explanation as to whether or not the negative results of this study represents a true failure of misonidazole to improve control in patients with the type of cerebral tumours treated. It seems more likely that the study was inadequate for some of the reasons listed above; and in particular the fractionation schedule and inadequate number of patients entered into the trial. It is to be hoped that the several new studies on patients with this tumour now in progress, will soon yield a definitive answer.

DESMETHYLMISONIDAZOLE

We have recently commenced early clinical studies with desmethylmisonidazole given as an oral suspension to volunteer patients. Interest in the drug arises from the observations that its acute LD_{50} in mice is greater than that of misonidazole but that it is as effective a hypoxic cell radiosensitiser (7-9). Observations in dogs have shown it to have a shorter $t_{\frac{1}{2}}$ than misonidazole; its penetration into CSF and central nervous tissue was reduced because of its decreased lipophilicity, whilst penetration into a variety of spontaneous canine tumours remained adequate (23).

Our preliminary studies are confirming the findings report in dogs. The plasma $t_{\frac{1}{2}}$ in patients is around 6h as compared with approximately double that previously seen for misonidazole. In one patient with a squamous carcinoma of a pilonidal sinus multiple tumour biopsies after a dose of desmethylmisonidazole of $1g/m^2$ showed peak tumour concentrations of the drug of 119, 130 and 133% of the plasma levels at 2, 3 and 4h respectively after oral administration (patient D, Fig.1). The plasma $t_{\frac{1}{2}}$ in this patient was 6.3h (5.9 - 6.7).

In another patient with a cerebral astrocytoma (patient C, Fig.1), biopsies were taken of normal brain and tumour 3h after an oral dose of $1g/m^2$ of the drug. Tumour concentration was 82% of that of plasma whilst that in brain was only 19%. A sample of CSF contained only 18% of the plasma concentration. The levels in normal brain and CSF confirmed that dog data (23), and the likely effect of low lipophilicity on drug penetration across the blood brain barrier. It is of considerable interest, however, that brain tumour penetration remains

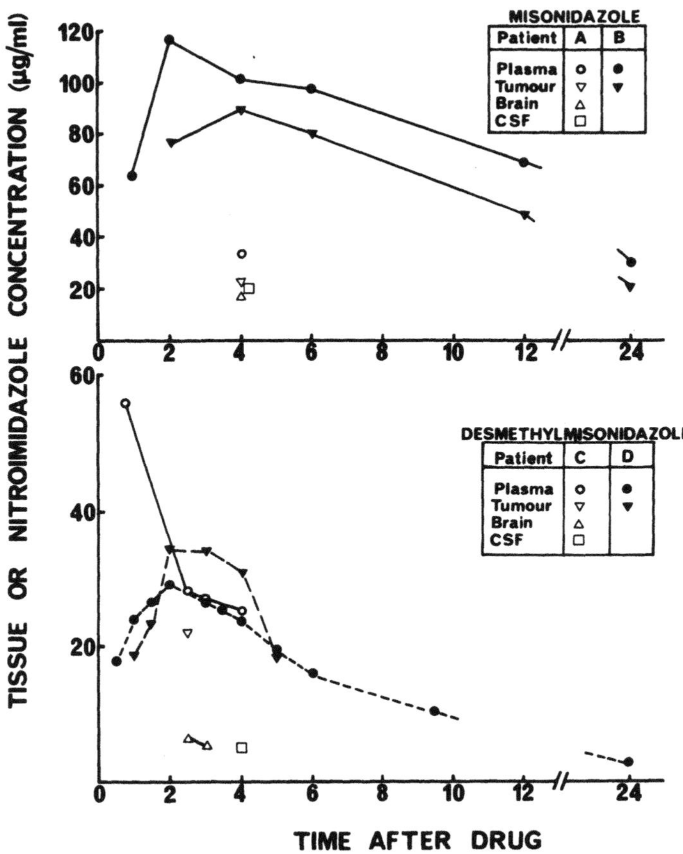

Fig.1.

Tissue concentrations of drug in four patients receiving either misonidazole or desmethylmisonidazole :

patient A (glioma) given $1g/m^2$ misonidazole

patient B (breast carcinoma) given $3g/m^2$ misonidazole

patient C (glioma) given $1g/m^2$ desmethylmisonidazole

patient D (squamous carcinoma) given $1g/m^2$ desmethylmisonidazole high, presumably because the barrier is no longer intact at this site. We are currently extending this series of biopsy data.

Details of two other patients (A,B) given misonidazole are also presented in Fig.1 (upper panel) for comparison.

If the incidence of neurotoxicity is related to lipophilicity as has been postulated(23),and also the AUC of the plasma drug concentrations (24), then desmethylmisonidazole is less likely to produce neuropathy than misonidazole. We are currently testing this hypothesis clinically. Of course the real criteria of value of the drug will be if its therapeutic gain factor is increased over that of misonidazole. This will take much longer to determine as we have yet to prove a clinical advantage for misonidazole in a major clinical trial situation. Nonetheless, neurotoxicity is the major dose limiting complication for misonidazole and it may well be possible to use higher doses of desmethylmisonidazole without a comparable increase in its toxicity.

SUMMARY AND CONCLUSIONS

A review is presented of clinical data obtained from the author's studies with the hypoxic cell radiosensitiser, misonidazole. It has been shown that the plasma pharmacokinetics of the drug may be modified by prior induction of hepatic microsomal enzymes with either phenytoin or phenobarbitone. Biopsy of brain and other tumour samples has shown good penetration of the drug at 3 - 5 h after oral administration. Similar drug concentrations are also seen in normal brain and CSF. Preliminary results of a similar study with desmethylmisonidazole are reported.

The results of a randomised trial of radiotherapy with or without misonidazole in a total of 55 patients with grade 3 and 4 cerebral astrocytomes are presented. No improvement in survival was seen as a result of the radiosensitisation and the significance of this observation is discussed.

This paper has only reviewed limited aspects of the clinical studies with two hypoxic cell radiosensitizers, one of which is a metabolite of the other. Many centres are actively involved in pursuing such studies and it is likely that definitive answers, at least for misonidazole, should be available within the next year or so. Clinical studies take a long time to implement; complete; analyse and assess the significance of their results. It may be that the improvements in results from the use of misonidazole will not be as good as early expectations. If this is the case, it is important not to let this influence enthusiasm for further trials. Many new drugs are being developed and tested experimentally. Desmethylmisonidazole is only one such agent which has now reached clinical investigation but there may be others even more promising. It is therefore most important that the next generation of clinical radiosensitiser trials is only entered upon after very vigorous review of all the newly available drugs. We owe this both to the discipline of the science and to the well-being of our patients.

Acknowledgements

I wish to thank numerous colleagues for their participation in the various aspects of this work and in particular Drs. R. Grant, C. R. Wiltshire and P. Workman. Misonidazole and desmethylmisonidazole was provided by Dr. I. Lenox-Smith (Roche Limited, Welwyn, England).

REFERENCES

1. Hypoxic Cell Sensitisers in Radiobiology and Radiotherapy. 8th L. H. Gray Conference, Cambridge 5th-9th Sept. 1977. Br. J. Cancer 38:Suppl.II (1978).
2. Combined Modality Cancer Treatment. Radiation Sensitiser and Protectors Meeting. Key Biscayne, Florida. Oct. 3rd-6th, 1979. Cancer Clinical Trials 3:in press (1980).
3. C. R. Wiltshire, P. Workman, J. V. Watson and N. M. Bleehen, Clinical Studies with Misonidazole. Br.J.Cancer 37:Suppl.III, Br.J.Cancer 37:Suppl.III, 286 (1978).
4. P. Workman, C. R. Wiltshire, P. N. Plowman and N. M. Bleehen, Monitoring Salivary Misonidazole in Man : A possible alternative to plasma monitoring. Br.J.Cancer 38:709 (1978).
5. P. Workman, N. M. Bleehen and C. R. Wiltshire, Phenytoin shortens the half-life of the hypoxic cell radiosensitiser misonidazole in man : Implications for possible reduced toxicity. Br.J.Cancer 41: 307 (1980).
6. N. M. Bleehen, The Cambridge Glioma Trial of Misonidazole and Radiation Therapy with Associated Pharmacokinetic Studies. Cancer Clinical Trials 3:in press (1980).
7. G. E. Adams, I. R. Flockhart, C. E. Smithen, I. J. Stratford, P. Wardman and M. E. Watts, Electron affinic Sensitisation, VII : A correlation between structures, one-electron reduction potential and efficiencies of nitroimidazoles as hypoxic cell radiosensitisers. Radiat. Res. 67:9 (1976).
8. I. R. Flockhart, P. W. Sheldon, I. J. Stratford and M. J. Watts, A Metabolite of the 2-nitroimidazole Misonidazole with Radiosensitising properties. Int. J. Radiat. Biol. 34:91 (1978).
9. J. M. Brown, N. Y. Yu and P. Workman, Pharmacokinetic considerations in testing Hypoxic Cell Radiosensitisers in Mouse Tumours. Br. J. Cancer 39:310 (1979).
10. P. Workman, C. J. Little, T. R. Marten, A. D. Dale, R. J. Ruane, I. R. Flockhart and N. M. Bleehen, Estimation of the Hypoxic Cell Sensitiser Misonidazole and its O-demethylated metabolite in

biological materials by reversed-phase high performance liquid chromatography. J.Chromatog. 145:507 (1978).
11. P. Workman, Effects of pretreatment with Phenobarbitone and Phenytoin on the Pharmacokinetics and Toxicity of Misonidazole in Mice. Br.J.Cancer 40:335 (1979).
12. S. Dische, M. I. Saunders, P. Anderson, R. C. Urtasun, K. H. Karcher, H. D. Kogelnik, N. M. Bleehen, T. L. Phillips, T. H. Wasserman, The Neurotoxicity of Misonidazole : pooling data from five centres. Br.J.Radiol.51:1023 (1978).
13. T. H. Wasserman, T. L. Phillips, G. Van Raalte, R. Urtasun, J. Partington, D. Koziol, J. G. Schwade, D. Gangji and J. M. Strong, The Neurotoxicity of Misonidazole : potential modifying role of phenytoin sodium and dexamethasone. Br.J.Radiol. 53:172 (1980).
14. P. Workman, The Neurotoxicity of Misonidazole : potential modifying role of dexamethasone. Br.J.Radiol. 53:736 (1980).
15. D. Ash, M. R. Smith and R. Bugden, The distribution of Misonidazole in Human Tumours and Normal Tissues. Br.J.Cancer 39:503-509, 1979.
16. S. Dische, M. I. Saunders, M. E. Lee, G. E. Adams and I. R. Flockhart, Clinical Testing of the Radiosensitiser Ro 07-0582. Experience with multiple doses. Br.J.Cancer 35:567 (1977).
17. R. C. Urtasun, J. D. R. Miller, V. Frunchak, D. Koziol, P. R. Band, J. D. Chapman and M. L. Feldstein, Radiotherapy Pilot Trials with Sensitisers of Hypoxic Cells : Metronidazole in supratentorial glioblastomas. Br.J.Radiol. 50:602 (1977).
18. S. Hornsey and A. White, Isoeffect Curve for Radiation Myelopathy. Br.J.Radiol. 53:168 (1980).
19. A. J. Van der Kogel, Late Effects of Radiation on the Spinal Cord. Dose effect relationships and pathogenesis. T.N.O., Rijswijk, Holland (1979).
20. J. Denekamp, N. J. McNally and J.F. Fowler, Misonidazole in Fractionated Radiotherapy : Little and often? Br.J.Radiol. : in press.
21. J. F. Fowler, G. E. Adams and J. Denekamp, Radiosensitisers of Hypoxic Cells in Solid Tumours. Cancer Treat. Rev. 3:227 (1976).
22. J. M. Yuhas, Misonidazole Enhancement of Acute and Late Radiation Injury to Rat Spinal Cord. Br.J.Cancer 40:161 (1979).
23. R. A. S. White and P. Workman, Pharmacokinetic and Tumour-penetration Properties of the Hypoxic Cell Radiosensitiser Desmethylmisonidazole (Ro 05-9963) in Dogs. Br.J.Cancer 41: 268 (1980).
24. M. I. S. Saunders, S. Dische, P. Anderson and I. R. Flockhart, The Neurotoxicity of Misonidazole and its relationship to dose, half-life and concentration in serum. Br.J.Cancer 37 (Suppl.III): 268 (1978).

PARTICIPANTS

(C - conference only)

D. AMADORI
Ospedali Civile-Oncologia
Forli, Italy.

L. BABINI
Istituto di Radiologica
e de Radio "L. Galvani"
Universita di Bologna
Policlinico "S. Orsola"
Bologna, Italy.

R. BADIELLO
Laboratorio FRAE
CNR
Via Çastaguoli 3
Bologna, Italy.

R. BALDUCCI
Cattedra di Chimica Generale
ed Inorganica
Facoltà Farmacia
Università di Bologna
Bologna, Italy.

S. BAR-POLLAK
National Institute of Oncology
H-1118 Budapest
Otthon u. 27
Hungary

A. C. BENINI
Istituto di scienze Chimiche
Facoltà di Farmacia
Università di Bologna
Via S. Donato 15
Bologna, Italy.

T. BERGAMI
Divisione di Radioterapia
Ospedale "M. Malpighi"
V. le Albertoni 15
Bologna, Italy.

G. BERTOCCHI (C)
Istituto Chimico "G. Ciamician"
Universita di Bologna
Via Selmi 2
Bologna 40126
Italy.

M. BIGNARDI
Ospedale del Circolo Varese
Via Borri 57
Varese 21100, Italy.

R. H. BISBY
Cockroft Building
University of Salford
Salford M5 4WT, UK.

A. BLOTTA
Divisione di Radioterapia
Ospedale "M. Malpighi"
V. le Albertoni 15
Bologna, Italy.

M. BOTTCHER
Universität Hamburg
D-2059 Büchen
Weisenweg, Germany.

A. BOZZI
Istituto di Chimica Biologica
Università degli Studi
P. le Aldo Moro
Rome, Italy.

L. CIONINI
Istituto di Radiologia
Policlinico
Università di Firenze
Firenze 50134, Italy.

C. CLARKE
Institute of Cancer Research
Clifton Avenue
Sutton, Surrey, UK.

L.A. COHEN (C)
Department of Health
National Institutes of Health
Building 4, Room 330
Bethesda
Maryland 20014, USA.

F. COUCOURDE
Istituto Tumori
Fondazione Pascale
Napoli 80100, Italy.

D. COVIC
Central Institute for Tumours
and Allied Diseases Zagreb
Ilica 197
Zagreb 41000, Yugoslavia.

S. DAL FIOR
Reparto di Radioterapia
Ospedale Civile
Vicenza 36100, Italy.

M. D'ANGELANTONIO
Cattedra di Chimica Generale
Istituto di Scienze Chimiche
Facoltà di Farmacia
Via S. Donato 15
Bologna, Italy.

S. DE GARIS
F. Hoffmann-La Roche
CH-4002
Basel, Switzerland.

G. DE GIULI
Presidente, SARO-SISM
Istituto di Radiologica Medica
Università degli Studi
V. le Morgagni-Careggi
Firenze 50134, Italy.

D. DE MARIA
Istituto di Radiologia
Università degli Studi
Modena 41100, Italy.

E. EMILIANI
Istituto di Radioterapia
"L. Galvani"
Policlinico "S. Orsola"
Via Massarenti 9
Bologna 40126, Italy.

P. ENGSTROM (C)
AB Leo Rhodia
Fack, S-251 00
Helsingborg, Sweden.

M.G. FABRINI
Istituto di Radiologia
Università di Pisa
P.za Gorgona 21
Marina di Pisa
Pisa 56100, Italy.

F. FASANO
Divisione di Radioterapia
Ospedale "M. Malpighi"
Bologna, Italy.

F. FORTUNATO
Fisica Sanitaria
Fondazione Sen. Pascale
Via M. Semmola 1
Napoli 80100, Italy.

F.C. GARDINI
Casa di Cura
Villa Torri
V. le Filopanti 12
Bologna 40126, Italy.

P. GOLDMAN
Beth Israel Hospital
330 Brookline Avenue
Boston
Massachusetts 02215
USA.

G. GRASHEV
Tumor Radiobiology Laboratory
The National Center of Oncology
Plovdivsko Pole 6
Sofia 11-56, Bulgaria.

M.G. GRILLI
Istituto di Scienze Chimiche
Facoltà di Farmacia
Università di Bologna
Via S. Donato 15
Bologna 40127, Italy.

H. HAKKOUSH
Radiotherapy Department
National Cancer Institute
Kars El Aini Street
Fom El Khaling
Cairo, Egypt.

C. HOUEE-LEVIN
Laboratoire de Chimie-Physique
Université Paris V
45 rue des Saints-Peres
75270 Paris Cedex 06
France.

W.R. INCH
The Ontario Cancer Foundation
London Clinic
Victoria Hospital
South Street
London, Ontario N6A 4G5
Canada.

A.M.M. JOKIPII (C)
Department of Serology
and Bacteriology
University of Helsinki
Helsinki, Finland.

L. JOKIPII (C)
Department of Serology
and Bacteriology
University of Helsinki
Helsinki, Finland.

J.S. KEINTHLY (C)
Cornell University Medical School
Department of Medicine
Division of International Medicine
1300 York Avenue
New York 10021, USA.

R.C. KNIGHT
Department of Applied Biology
North East London Polytechnic
Romford Road
London E15 4LZ, UK.

R.J. KNOX
Chemotherapy Research Unit
North East London Polytechnic
Romford Road
London E15 4LZ, UK.

F. LANCEWICZ
Istituto di Oncologia
"Felice Addarii" di Bologna
Via E. Duse 7
Bologna 40100, Italy.

G.C. LANCINI
Lepetit
Via Durando 38
Milano 20158, Italy.

M.A. LELLI
Viale dei Mille 72
Cesenatico
Forli 47100, Italy.

I. LENOX-SMITH
Roche Products Ltd.
P.O.Box 8
Welwyn Garden City
Herts AL7 3AY, UK.

L. LEONARDI
Divisione di Radioterapia
Ospedale "M. Malpighi"
V. le Albertoni 15
Bologna 40100, Italy.

H. MA
Institute of Cancer Research
Clifton Avenue
Sutton, Surrey SM2 5PX, UK.

G. MAGNONI
Arcispedale S. Anna
Istituto di Radiologia
Via G. da Carpi 6
Ferrara 44100, Italy.

L. MAHOOD
Department of Biochemistry
Brunel University
Uxbridge, Middlesex, UK.

M. MAKROPOULOU
Institute of Medicine and Pharmacology
Department of Biophysics
Bd. Petru Groza 8
Bucharest, Romania.

C. MALTONI
Istituto di Oncologia
"F. Addarii"
Policlinico S. Orsola
Bologna, Italy.

T.R. MARTEN
Roche Products Ltd.
Welwyn Garden City
Herts AL7 3AY, UK.

F. MONDINI
Istituto di Scienze Chimiche
Facoltà di Farmacia
Università di Bologna
Via S. Donato 15
Bologna 40127, Italy.

C. NERVI
Department of Radiation Oncology
Clinica Villa Flaminia
Via L. Bodio 58
Roma 00100, Italy.

I. NILSSON-EHLE (C)
Hospital of Lund
Lund, Sweden.

C.E. NORD (C)
National Bacteriological Lab.
Stockholm
S-105 21 Stockholm, Sweden.

L. OLIVETTI
Spedali Civili di Brescia
Istituto del Radio
"O. Alberti"
Brescia 25100, Italy.

P. PACINI
Servizio di Radioterapia
Arcispedale S. Maria Nuova
V. le Morgagni-careggi
Firenze 50134, Italy.

M. PASINI
Istituto di Scienze Chimiche
Facoltà di Farmacia
Università di Bologna
Bologna 40127, Italy.

V.T. PASQUINELLI
Istituto di Radioterapia
Ospedale S. Orsola
Università
Via Massarenti 9
Bologna 40126, Italy.

C. PISONI
Farmitalia - Carlo Erba
Via Imbonati 24
Milano 20159, Italy.

G. PIZZI
Divisione di Radioterapia
Ospedale Civile
Padova 35100, Italy.

M. QUINTILLIANI
Laboratorio di Biologia
CNEN
Casaccia
Roma, Italy.

A.V. REYNOLDS (C)
The School of Pharmacy
University of London
Department of Pharmaceutics
29/39 Brunswick Square
London WC1N 1AX, England.

T. RIPA (C)
Department of Clinical
Bacteriology
Lanssjukhuset
S-301 85 Halmstad, Sweden.

P.T. RISSANEN (C)
Radiotherapy Clinic
University Central Hospital
Haartmaninkatu 4
00290 Helsinki 29
Finland.

L. ROCCHI
Ospedale Civile Cesena
Centro Tumori
Cesena
Forli, Italy.

D. ROWLEY
Department of Applied Biology
North East London Polytechnic
Romford Road
London E15 ALZ, UK.

D. RUBIN
National Heart, Lung and Blood
Institute
Building 31, Room 5A03
9000 Rockville Pike
Bethesda
Maryland 20205, USA.

G. SEMERANO
Via Stellini 8
Padova 35100, Italy.

R.A. SMITH
Department of Biology
Medical and Biological Sciences
Building
Bassett Crescent East
Southampton SO9 3TU, UK.

T. SOYLEMEZ
Institut fur Biophysics
und Elektronenmikroskopie
der Universität Düsseldorf
D-4000 Düsseldorf
Mooren Strasse 5, Germany.

M. SPANO
Dosimetria Biofisica
CNEN
Casaccia
Via Anguillarese Km. 1300
Roma 00100, Italy.

D. SPINELLI
Cattedra di Chimica Organica
Facoltà di Farmacia
Università di Bologna
Via Zanolini 4
Bologna 40216, Italy.

R. SRIDHAR
Biomembrane Research Lab.
825 Northeast 13th Street
Oklahoma City
Oklahoma 73104, USA.

R. STAGNI
Istituto di Scienze Chimiche
Facolta di Farmacia
Universita di Bologna
Via S. Donato 15
Bologna 40127, Italy.

M.R.L. STRATFORD
CRC Gray Laboratory
Mount Vernon Hospital
Northwood
Middlesex HA6 2RN, UK.

J. SUWINSKI (C)
Institute of Organic Chemistry
and Technology
Silesian Polytechnical Univ.
Gliwice 44-100, Poland.

M. TAMBA
Laboratorio FRAE
CNR
Via Castaguoli 3
Bologna 40126, Italy.

P. TAMULEVICIUS
Istitut fur Strahlen Biol
Universität-Klinikum Essen
45 Essen
Hufelanstrasse 55
W. Germany.

E. TEKAVCIC
Institute of Oncology
61000 Ljubljana
Vrazov TRG 4, Yugoslavia.

R. TEMPLETON
May & Baker Ltd
Pharmacology
Dagenham
Essex RM1 OXS, UK.

B. URSING (C)
Department of Infectious
Diseases
University Hospital
S-221 85 Lund, Sweden.

A.Y. WIJNMAALEN
Rotterdamsch Radio-Therap.
Institute
Groene Hilledijk
3075 EA Rotterdam, Holland.

R.L. WILLSON
Department of Biochemistry
Brunel University
Uxbridge
Middlesex UB8 2QH, UK.

E. WINKELMANN (C)
Hoechst AG
Pharma Synthese G 838
Postfach 80 03 20
6130 Frankfurt/Main 80
W. Germany.

S. YASHVIR
Jawaharlal Nehru University
School of Life Sciences
New Mehrauli Road
New Delhi - 110067, India.

A. YERUSHALMI
Weizmann Institute of Sciences
P.O.Box 26
Rehovot, Israel.

K.L. YEUNG
Texas Medical Centre
Houston
Texas 77030, USA.

INDEX

GPSR Compliance

The European Union's (EU) General Product Safety Regulation (GPSR) is a set of rules that requires consumer products to be safe and our obligations to ensure this.

If you have any concerns about our products, you can contact us on ProductSafety@springernature.com

In case Publisher is established outside the EU, the EU authorized representative is:

Springer Nature Customer Service Center GmbH
Europaplatz 3
69115 Heidelberg, Germany

Batch number: 10615115

Printed by Printforce, the Netherlands